A Primer on Compression in the Memory Hierarchy

Synthesis Lectures on Computer Architecture

Editor
Margaret Martonosi, *Princeton University*

Founding Editor Emeritus
Mark D. Hill, *University of Wisconsin, Madison*

Synthesis Lectures on Computer Architecture publishes 50- to 100-page publications on topics pertaining to the science and art of designing, analyzing, selecting and interconnecting hardware components to create computers that meet functional, performance and cost goals. The scope will largely follow the purview of premier computer architecture conferences, such as ISCA, HPCA, MICRO, and ASPLOS.

A Primer on Compression in the Memory Hierarchy

Somayeh Sardashti, Angelos Arelakis, Per Stenström, and David A. Wood

ISBN: 978-3-031-00623-4   paperback
ISBN: 978-3-031-01751-3   ebook

DOI 10.1007/978-3-031-01751-3

A Publication in the Springer series
SYNTHESIS LECTURES ON ADVANCES IN AUTOMOTIVE TECHNOLOGY

Lecture #36
Series Editor: Margaret Martonosi, *Princeton University*
Founding Editor Emeritus: Mark D. Hill, *University of Wisconsin, Madison*
Series ISSN
Print 1935-3235   Electronic 1935-3243

A Primer on Compression in the Memory Hierarchy

Somayeh Sardashti
University of Wisconsin, Madison

Angelos Arelakis
Chalmers University of Technology

Per Stenström
Chalmers University of Technology

David A. Wood
University of Wisconsin, Madison

SYNTHESIS LECTURES ON COMPUTER ARCHITECTURE #36

ABSTRACT

This synthesis lecture presents the current state-of-the-art in applying low-latency, lossless hardware compression algorithms to cache, memory, and the memory/cache link. There are many non-trivial challenges that must be addressed to make data compression work well in this context. First, since compressed data must be decompressed before it can be accessed, decompression latency ends up on the critical memory access path. This imposes a significant constraint on the choice of compression algorithms. Second, while conventional memory systems store fixed-size entities like data types, cache blocks, and memory pages, these entities will suddenly vary in size in a memory system that employs compression. Dealing with variable size entities in a memory system using compression has a significant impact on the way caches are organized and how to manage the resources in main memory. We systematically discuss solutions in the open literature to these problems.

Chapter 2 provides the foundations of data compression by first introducing the fundamental concept of value locality. We then introduce a taxonomy of compression algorithms and show how previously proposed algorithms fit within that logical framework. Chapter 3 discusses the different ways that cache memory systems can employ compression, focusing on the trade-offs between latency, capacity, and complexity of alternative ways to compact compressed cache blocks. Chapter 4 discusses issues in applying data compression to main memory and Chapter 5 covers techniques for compressing data on the cache-to-memory links. This book should help a skilled memory system designer understand the fundamental challenges in applying compression to the memory hierarchy and introduce him/her to the state-of-the-art techniques in addressing them.

KEYWORDS

cache design, memory system, memory hierarchy, data compression, performance, energy efficiency

Contents

List of Figures

List of Tables

Preface

This primer is intended for readers who are interested in learning about the different ways that data compression can be applied to the computer memory hierarchy, including caches, main memory, and the links that connect them. This audience includes computing industry professionals and graduate students. We expect our readers to be familiar with the basics of computer architecture. Knowing the details of out-of-order execution is unnecessary, but readers should be comfortable with the basics of cache and memory hierarchy design.

This primer's primary goal is to provide readers with a basic understanding of the key challenges and opportunities in applying data compression to the memory hierarchy. We address the complementary issues of data compression and data compaction, presenting high-level concepts and how they manifest in previously proposed designs. We introduce a taxonomy to help understand which compression algorithms are most applicable to different levels of the memory hierarchy. We use another taxonomy to classify different ways of compacting variable-size compressed blocks into memory structures that are classically designed to hold fixed-size blocks.

A secondary goal of this primer is to make readers aware of the opportunity that data compression has to improve the performance, energy efficiency, and cost of future computer systems. We believe that technology trends are converging in such a way as to make data compression a compelling solution and hope this primer may help expedite its widespread adoption throughout the memory system. It is not a goal of this primer to cover all topics in depth, but rather to cover the basics and help readers identify which topics they may wish to pursue in greater depth.

Somayeh Sardashti, Angelos Arelakis, Per Stenström, and David A. Wood
December 2015

Acknowledgments

We owe many thanks for the help and support we have received during the development of this primer. We would like to thank Alaa Alameldeen, Martin Burtscher, Magnus Ekman, Aamer Jaleel, and Martin Thuresson for their helpful feedback on this work. While our reviewers provided great feedback, they may or may not agree with all of the final content of this primer.

We would also like to thank our former co-authors Alaa Alameldeen, Chloe Alverti, Fredrik Dahlgren, Magnus Ekman, Andre Seznec, and Martin Thuresson for their role in helping us understand the issues in applying data compression to the memory hierarchy. Without their contributions, this work would not have been possible.

This work was supported in part by the National Science Foundation (CCF-1218323, CNS-1302260, CCF-1438992, CCF-1533885). Professor Wood has a significant financial interest in AMD and Google. The views expressed herein are not necessarily those of the National Science Foundation, nor anyone other than the authors. This work was also supported in part by the CHAMPP project funded by the Swedish Research Council, the FP7 EUROSERVER project funded by the European Commission, and the SCHEME project funded by the Swedish Foundation for Strategic Research.

Somayeh thanks her colleagues and coauthors for sharing their expertise, her husband, Dr. Hamid Reza Ghasemi, and parents, Aghdas Zeinali and Khosro Sardashti, for all their love, support, and encouragement.

Angelos thanks his coauthors of this book for the pleasant collaboration in this inspiring work, his parents Dimitris and Mary, and his sister Stella for encouraging him to always follow his dreams, and his wife Christina for always being on his side and making his life joyful and for her love.

Per thanks all his collaborators in the past and foremost the great collaboration with the coauthors of this book for an inspiring and very enjoyable mission. Most importantly, he thanks his wife Carina and their daughter Sofia for all the love and support they provide.

David thanks his coauthors for putting up with his deadline-challenged work style, his parents Roger and Ann Wood for inspiring him to be a second-generation Computer Sciences professor, and Jane, Alex, and Zach for helping him remember what life is all about.

Somayeh Sardashti, Angelos Arelakis, Per Stenström, and David A. Wood
December 2015

CHAPTER 1

Introduction

The memory hierarchy of a computer is a critical system component, impacting the system's overall performance, energy consumption, and cost. As processors have become faster, the memory hierarchy has become deeper (i.e., more levels) to help bridge the speed, energy, and bandwidth gaps between processors and relatively slower (and larger) main memories. Today's processors are typically clocked at a few GHz, while a main-memory access may take many tens of nanoseconds. Since high-performance processors may execute many instructions per cycle, a single main-memory access may consume the same time as a few hundred instructions. Similarly, accessing an off-chip memory location consumes one or more orders of magnitude greater energy than an access to an on-chip memory location. Finally, higher (i.e., smaller) levels of the memory hierarchy typically have significantly higher bandwidth than lower (i.e., larger) levels, resulting in a significant bandwidth gap.

Cache memories play a critical role in helping bridge these gaps, exploiting spatial and temporal locality to reduce the number of references to lower levels of the hierarchy. However, modern memory hierarchies frequently have three and even four levels of cache, resulting in a significant fraction of the cost of a system. For example, the Oracle SPARC M7 processor has a three-level cache hierarchy that accounts for roughly 50% of the total die area (and thus approximately 50% of the manufacturing cost). Thus, it is important to seek architectural techniques that can improve the efficacy of cache memories, by increasing their effective capacity and bandwidth. Similarly, while the cost of an individual DRAM chip is quite low, the sheer size of main memories, especially in server systems, makes memory cost a significant fraction of the total system cost. In addition, the cost of high-speed pins makes memory bandwidth a limited and expensive resource. Thus, architectural techniques that improve the effective capacity and bandwidth of main memory are also important.

This work explores the use of *data compression* to improve the efficiency of the memory hierarchy. Data compression has long been used in a diverse set of applications to store and communicate information using a smaller number of bits. Data compression has been used very effectively in long-distance communication channels, where the latency of compressing the original *data message* and decompressing the resulting *code words* represent a small fraction of the end-to-end latency. Data compression has also been widely used with media data, such as voice and video, which are amenable to *lossy*, rather than *lossless*, compression. Lossy compression is based on a quality-of-service model, where semantic knowledge can help determine which data are critical to achieving a higher quality of service. Lossy compression can greatly reduce the number of bits needed to represent media data, but cannot in general exactly reproduce the original quality. Lossless compression, on the other hand, requires that data be compressed and decompressed with no information loss compared to the originally compressed data. When applying data compression to the memory hierarchy, architects must in general seek out lossless compression algorithms that incur low latencies.

This synthesis lecture presents the current state-of-the-art in applying low-latency, lossless hardware compression algorithms to cache, memory, and the memory/cache link. There are many non-trivial challenges that must be addressed to make data compression work well in this context. First, since compressed data must be decompressed before it can be accessed, decompression latency ends up on the critical memory access path. This imposes a significant constraint on the choice of compression algorithms. Second, while conventional memory systems store fixed-size entities like data types, cache blocks, and memory pages, these entities will suddenly vary in size in a memory system that employs compression. Dealing with variable size entities in a memory system using compression has a significant impact on the way caches are organized and how to manage the resources in main memory. We systematically discuss solutions in the open literature to these problems.

Chapter 2 provides the foundations of data compression by first introducing the fundamental concept of value locality. We then introduce a taxonomy of compression algorithms and show how previously proposed algorithms fit within that logical framework. Chapter 3 discusses the different ways that cache memory systems can employ compression, focusing on the trade-offs between latency, capacity, and complexity of alternative ways to compact compressed cache blocks. Chapter 4 discusses issues in applying data compression to main memory and Chapter 5 covers techniques for compressing data on the cache-to-memory links. Finally, Chapter 6 presents concluding remarks.

CHAPTER 2

Compression Algorithms

In information theory, the entropy of a source input is the amount of information contained in that data [126]. Entropy determines the number of bits needed to optimally represent the original source data. Therefore, entropy sets an upper bound on the potential for compression. Low entropy suggests that data can be represented with fewer bits. Although computer designers try to use efficient coding for different data types, the memory footprints of many applications still have low entropy.

Compression algorithms compress a *data message* into a series of *code words* by exploiting the low entropy in the data. These algorithms map a data message to compressed code words by operating on the data message as a sequence of input symbols, e.g., bits, bytes, or words. In this section, we first establish the information theoretic foundations, then introduce a taxonomy of algorithms, and classify different compression algorithms. We then introduce the main metrics to evaluate the success of a given compression algorithm.

2.1 VALUE LOCALITY

Computers access and process data in chunks of particular sizes depending on the data types forming the values. For example, some accesses are for 64-bit floating-numbers whereas others are for 32-bit integers. During program execution, it is possible for a previously accessed value to be accessed again either in the same memory location or in another location [112]. We refer to the property that exactly the same value or a set of similar values are replicated across multiple memory locations, as *value locality* [113].

Conventional cache/memory hierarchies seek to exploit the principle of *reference locality, predicting that the same blocks will be frequently accessed*. Therefore, when programs that exhibit value locality run on systems with conventional cache/memory hierarchies, a subset of data values may be replicated and saved in several different memory/cache locations causing *value replication*.

A computational unit (or kernel) of a program typically generates a stream of references (addresses), as shown in the left part of Figure 2.1. It is well known that most programs exhibit *temporal and spatial reference locality*, as illustrated in the left part of Figure 2.1. *Temporal locality* says that recently accessed *addresses* tend to be accessed again soon. *Spatial reference locality* says that nearby *addresses* tend to be accessed soon. What is less well established, however, is that programs also tend to exhibit *value locality*. Like reference locality, value locality can be further classified into *temporal and spatial value locality*, but it is a property of *data values*, not *addresses*. *Temporal value locality* says that for any given data value accessed, the same value is likely to

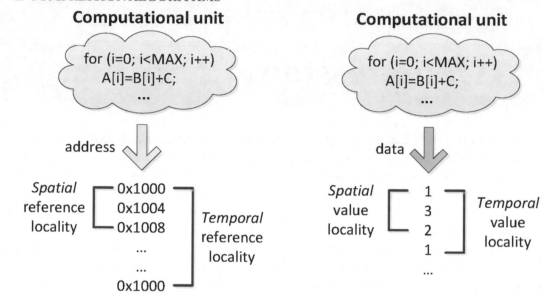

Figure 2.1: Notions of reference and value locality.

be accessed again. *Spatial value locality* says that for any given data value accessed, numerically similar or "nearby" values are likely to be accessed [113]. A high value locality suggests that many cache/memory locations contain the same or similar value. Hence, a memory system that would be able to keep a single copy of each distinct value could offer high compressibility. Studies have shown that such a memory system could offer significant compressibility; up to 64X have been demonstrated for some SPEC 2006 applications [113]. While such a memory system is hard to conceive, we will see that compression algorithms build on the value locality property. This is equally applicable to compression in caches, in memory, as well as on the link between the memory and the caches.

State-of-the-art compression approaches focus more often on temporal value locality than spatial value locality. For example, temporal value locality is exploited by approaches that try to eliminate:

- The value zero [23, 37, 52, 59, 60, 115].

- A small set of frequent values established statically [38–41] or dynamically [18, 38].

- A large set of common values by capturing the unique instances of them in a dictionary, which is updated dynamically. The occurring values are replaced by pointers to dictionary entries and in some approaches by a combination of pointers and matched length. The most common dictionary-based compression algorithms are the Lempel-Ziv variations [27, 28].

- A large set of common values by encoding them with variable-length codewords based on their frequency of occurrence: short codewords for the most frequent values and longer codewords for the least frequent ones. Huffman coding [24] and Arithmetic encoding [26] are the most common statistical compression algorithms.

On the other hand, spatial value locality is mainly exploited by approaches that aim at eliminating value redundancy when values are numerically similar, thus creating value clusters. For this reason, spatial value locality is sometimes referred to as *clustered value locality* [114]. A cluster of values can be compressed by selecting one value—the *base value*—and encoding values by taking the difference between them and the base value [15, 19, 116, 117]. A special case occurs when the base value is zero, sometimes called *narrow value locality,* which leverages the well-known observation that small integers are quite common [123]. Examples of compression algorithms that build on narrow value locality are significance-based compression [18, 20, 120].

Fundamental to exploitation of value locality is the granularity of individual values. As the grain size increases, value locality is expected to go down since the number of unique possible values increases exponentially with the number of bits. On the other hand, it is more beneficial to encode large-grain values than small-grain values as the overhead of the encoding goes down as the grain size increases. For this reason, many approaches consider fixed-size fine-grain values of 4–8 bytes, which match common data types (e.g., 32- and 64-bit integers), others treat the input as a byte stream, and a few consider coarser value granularities, such as an entire 64-byte cache block. The compression schemes that fall into the latter category try either to eliminate blocks of the value zero (known as null blocks) [37, 59, 60] or take a different approach called *deduplication* [115, 118, 119]. Block deduplication seeks to eliminate the occurrence of blocks with the same content or "value."

Overall, many of these techniques form the foundation for several compression algorithms used in the memory hierarchy, as we will discuss in the next section.

2.2   COMPRESSION ALGORITHM TAXONOMY

This section introduces a taxonomy that will help classify the differences between the various loss-less compression algorithms that have been proposed for use in the memory hierarchy. In lossless algorithms, decompression can exactly recover the original data, while with lossy algorithms only an approximation of the original data can be recovered. Lossy algorithms are mostly used in voice and image compression where lost data does not adversely affect their usefulness. On the other hand, compression algorithms used in the memory hierarchy must, in general, be lossless since any single memory bit loss or change may affect the validity of a program's results.

Table 2.1 presents our taxonomy, including showing where a collection of well-known compression algorithms fit within it. For each algorithm, we classify it as:

- General purpose vs. special purpose

- Static vs. dynamic

Table 2.1: Compression algorithms taxonomy

		Temporal-Value Based	Spatial-Value Based
General-Purpose	**Static**	Static Huffman Coding [24] FVC [38] Null [37]	Significance-Based Address Compression [43][44] Null [37]
	Dynamic	Run-Length Encoding Lempel-Ziv [27][28] Dynamic Huffman Coding [25] C-PACK [18]	FPC [20] BDI [19] C-PACK [18]
Special-Purpose	**Static**	Instruction Compression [1]-[10]	—
	Dynamic	Floating-point Compression [12]-[16]	Floating-point Compression [127]

- Temporal-value based vs. spatial-value based

General-Purpose versus Special-Purpose: General-purpose algorithms target compressing data messages independent of their underlying data types or semantics. Many existing algorithms fall in this category, including BZIP2, UNIX gzip, and most algorithms used in compressed caches or memory. Conversely, specialized compression algorithms optimize for specific data types, exploiting the semantic knowledge of the data being compressed. Image/video compression and texture compression in GPUs [17] are good examples of specialized compression. By exploiting semantic content, special-purpose techniques tend to achieve higher compressibility for certain data types, including instructions and floating-point data. However, as data of various data types are stored and processed in computing systems, none of these special-purpose techniques always achieve the best compressibility. A first attempt to select the most appropriate compression method among different ones is done by HyComp [127], which proposes a hybrid compression method tailored for caches.

Static vs. Dynamic: Static compression algorithms provide a fixed mapping from the input data message to output code words [29]. They map a sequence of input bytes to the same set of code words every time that sequence appears in an input data message. Purely static algorithms use the same (static) mapping for all data messages. Pseudo-static algorithms use the same mapping for a given data message or set of messages, but may change the mappings for later messages. For example, some algorithms use two passes on the input data: one pass to determine the mapping, and a second pass for compression. Others change the mapping less frequently. For example, Semi-adaptive Huffman encoding [29] periodically measures the statistics of value frequency before updating the mapping, but then uses the same mapping for many messages [46]. On the other hand, dynamic algorithms require no previous knowledge of the data input and construct a mapping on the fly. For example, dictionary-based algorithms construct a dictionary of symbols (e.g., data values) on the fly and replace previously seen values with an index that identifies the corresponding dictionary entry. Deduplication is similar, but seeks to replace data messages (i.e., cache blocks) containing identical content with pointers to a unique instance of the block. Compression algorithms may also use a combination of static and dynamic sub-algorithms, such as a static mapping for very frequent values (e.g., zero) combined with a dictionary.

Temporal-Value Based vs. Spatial-Value Based: Compression algorithms can be classified into temporal-value based and spatial-value based algorithms. Temporal-value based algorithms seek to encode frequently occurring values (or sequences of values) using a smaller code word. For example, Frequent Value Compression [38] seeks to exploit the recurrence of a small number of distinct values that tend to be repeated very frequently (e.g., the values zero and one). Among frequently repeated values, zero is particular common, including a high occurrence of null blocks in many workloads. For example, Ekman and Stenström [59] show that 30% of 64-byte blocks are null in SPEC2000 and some server applications, on average. More recently Tian et al. [115] show that the fraction of null 64-byte blocks in SPEC2006 is 15%, on average. Null blocks tend to be more common in main memory than caches, due to zero padding at the end of pages and newly initialized (i.e., zeroed) pages. Dusser et al. propose a simple zero-block detection algorithm [37], which only takes one cycle to decompress.

Spatial-value based algorithms, on the other hand, are based on the observation that many values tend to be clustered around a common base value and can be encoded using this smaller difference. For example, Base-Delta-Immediate compression (BDI) [19] encodes integer values using the difference from one or more base values. Narrow-value-based algorithms exploit the special case where the base value is zero; that is, they exploit the frequency of small (or narrow) integers that do not require the full space allocated for them. For example, small integer values are sign-bit extended into 32-bit or 64-bit blocks, while all the information is contained in the least-significant bits. Frequent Pattern Compression [57], for example, replaces 32-bit words holding small values with a three-bit code and an eight-bit value. Note that significance-based compression also works well for zero words and null blocks, as zeros are small integer values. Significance-based algorithms have also been used to remove the redundant information in the high-order bits of addresses transferred between processor and memory [43, 44]. These schemes cache the high-order bits of addresses and only transfer the low-order address bits, as well as small indices (in place of the high-order address bits).

2.3   CLASSIFICATION OF COMPRESSION ALGORITHMS

In this section we present some popular compression algorithms that are frequently used for compressing data in the memory hierarchy. We also explain how they fit in our taxonomy.

2.3.1   RUN-LENGTH ENCODING

Run-length encoding (RLE) is arguably the simplest, and oldest, form of data compression. RLE encodes a contiguous sequence of identical input symbols with a single copy of the symbol and a count of the number of times it appears in the sequence. RLE works because contiguous sequences occur quite frequently in computer programs. Zeros are very common because most operating systems initialize newly allocated pages to be zero. The ASCII character 0x20 is very common, because many programs use the "space" character to represent an uninitialized string value. While rarely used alone, RLE forms an important part of many compression algorithms.

2.3.2   LEMPEL-ZIV (LZ) CODING

Lempel-Ziv (LZ) coding and its derivatives [27, 28] are arguably the most popular lossless dynamic compression algorithms. They form the basis for many software compression algorithms as well as several hardware implementations. LZ77 [27] manages the dictionary using a sliding window of previous values (symbols). The encoder tries to make the longest possible match of a value sequence in that window and encodes it using a 3-tuple *<offset, length, first unmatched value>*. The offset is the pointer to the dictionary, while the length corresponds to the matching amount (e.g., measured in symbols). As LZ77 tries to match the longest prefix in a value sequence, it exploits spatial value locality. On the other hand, LZ78 [28] compresses data by exploiting temporal value locality and building a dictionary on the fly of values that it has previously seen. A repeated (but not necessarily contiguous) symbol in the data message is replaced in the code word with a reference to the dictionary. If a match does not exist, it adds the new symbol to the dictionary. Like all dictionary algorithms, the LZ algorithms tend to work better for larger data messages (i.e., larger granularity) due to the greater opportunity for redundancy.

2.3.3   HUFFMAN CODING

Huffman algorithms represent more frequent symbols using shorter code words (i.e., fewer bits). Huffman coding derives a variable-length code table for each source symbol. It derives this table by building a binary tree, i.e., the Huffman tree, based on the probability or frequency of occurrence of each symbol in the data input. The tree is constructed bottom-up and left-to-right according to the frequency of occurrence of the symbols (their probabilities), starting from the least frequent ones. Figure 2.2 illustrates a stepwise construction of an example Huffman tree. In every step, the two least frequent values are selected and removed from the list; they are added to the tree and connected to a new internal node, whose probability is the sum of the probabilities of the previously added symbols. The new node is inserted to the list of symbols, which is sorted again. This process is repeated until one symbol is left in the list, with a probability equal to the sum of the initial probabilities. When the tree is ready, ones and zeros are assigned to the branches. A one (or a zero) can be assigned either to left branches or to right branches consistently (e.g., left branches are assigned to 1 and right branches are assigned to 0). Finally, the encoding is determined by traversing the tree from the root to each leaf. The derived codewords have the prefix property which stipulates that a valid codeword cannot be a prefix of another valid codeword. At the beginning (top left of Figure 2.2), the probabilities of the values (symbols) are sorted in increasing order. First, symbols E and D are removed from the list, inserted to the tree as leaves, and connected to a new (internal) node that is called ED and has a probability of $0.25(= P(E) + P(D) = 0.1 + 0.15 = 0.25)$. The node ED is inserted to the list in a position so that the list is still sorted in increasing order. This process is repeated until the list contains one node, the root of tree R. Zeros are assigned to the left branches and ones to the right branches of the tree. The generated Huffman encoding is depicted at the bottom left of Figure 2.2. In this example, the tree is binary (two nodes per parent node). However, the tree can be quaternary, or in general N-ary depending on how many child nodes each parent node may have. The tree structure affects the tree depth and thus the encoding as well as the decompression processing.

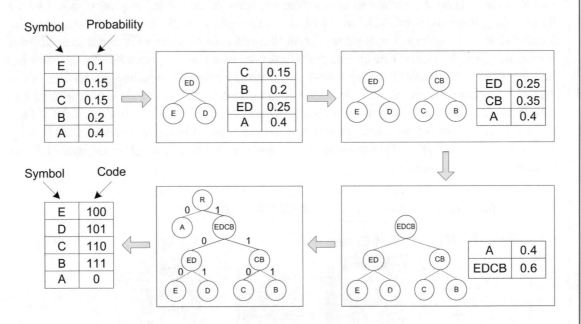

Figure 2.2: Example of building a Huffman tree.

Huffman coding can be generated statically or dynamically. Static Huffman coding [24] needs an extra pass on the input data to compute the probabilities (data pre-processing). Vitter [25] proposed dynamic Huffman coding that only requires one pass over the data; this way, the encoding is generated gradually during compression and both the compressor and decompressor maintain identical mapping trees. This requires compressed data to be decoded in the same order as they were encoded making dynamic Huffman encoding suitable only for communication channels and not for storage systems that require random access. There are also intermediate approaches between static and dynamic Huffman coding that we refer to as *semi-adaptive Huffman encoding* [29]. Semi-adaptive Huffman encoding requires two phases: One phase to establish the statistics of value frequency of the data to compress and generate the encoding, and a second phase to compress the data. As opposed to static Huffman coding that collects value statistics only once (e.g., with data pre-processing), semi-adaptive Huffman can collect value statistics and establish new encoding several times during the execution, thus it is expected to compress better than static. However, similarly to static coding it cannot compress newly generated unique symbols, unless a new encoding is constructed.

SC² [46] is a statistical compression scheme tailored specifically for caches that employs semi-adaptive Huffman encoding. There are two main processes (phases) that manage a statistical compression (Huffman-based) cache scheme: Sample and Compress, as illustrated in Figure 2.3. During the Sample process, the statistics of the most frequent uniquely-stored data values in the last-level cache (LLC) are established in a table structure, called Value Frequency Table (VFT), by recording every access to and from the LLC in a time-window. The authors observe that value locality tends to be quite stable over time for many workloads, thus the VFT needs to be updated rarely. The Sample process can be run periodically, or when data compressibility has dropped below a specific threshold. Because sampling is needed rarely, the encoding generation can be carried out using software routines. As data is assumed to be stored compressed only in the LLC, a Huffman Compressor (HuC) compresses all the blocks that are either inserted to the LLC from the main memory or written back from the upper-level caches. On the other hand, the Huffman Decompressor (HuD) decompresses every LLC block that is fetched to the upper-level caches or evicted to the main memory.

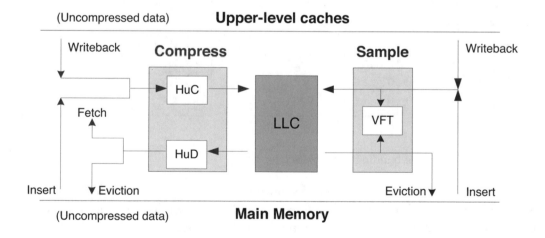

Figure 2.3: The statistical compression cache scheme (SC²) [46].

Table 2.2: Frequent pattern coding

Prefix	32-bit Data Word	Code Word Size (plus 3-bit prefix)
000	Zero Run (combines two words)	3 bits (runs up to 8 bytes)
001	4-bit sign extended	4 bits
010	8-bit sign extended	8 bits
011	16-bit sign extended	16 bits
100	16-bit sign extended	16 bits
101	Sign-extended halfwords	16 bits (two least significant bytes)
110	Repeated bytes	8 bits
111	Uncompressible word	32 bits

2.3.4 FREQUENT VALUE COMPRESSION (FVC)

Yang and Gupta [38] introduced the term *frequent value locality* and observed that for a running application at any execution point, a small number of distinct values occupy a large fraction of its memory footprint. They also found that the identity of the frequent values, which are fairly uniformly scattered across the memory, remains quite stable over the execution of a program. To find frequent values, they propose three different profiling approaches: one profiling run for each application before any main run, an initial profiling phase per application execution, and continuous profiling of a program during its execution. Thus, FVC is a temporal-value-based algorithm that, depending upon the profiling approach used, spans much of the spectrum between static and dynamic.

2.3.5 FREQUENT PATTERN COMPRESSION (FPC)

Frequent Pattern Compression (FPC) [57] is a general-purpose compression algorithm optimized for compressing small data messages. FPC exploits the fact that many values are narrow (e.g., small integers) and can be represented using a small number (e.g., 4–8) of bits, but are normally stored in full 32-bit or 64-bit words. Thus, FPC is a spatial-value-based algorithm. FPC compresses data messages on a word-by-word basis. FPC encodes common word patterns, compressing a 32-bit word to a 3-bit prefix and some data. Table 2.2 shows the different patterns corresponding to each prefix. FPC applies significance-based compression at word granularity (4 bytes), detecting and compressing a word to: 4 bits if 4-bit sign-extended; 8 bits if one-byte sign-extended or repeated byte; 16 bits if half-word sign-extended or half-word padded with a zero half-word; two half-words each a byte sign-extended. Figure 2.4 shows an example of FPC. In this example, the first word includes two half-words, each 1-byte sign-extended (e.g., 0xFF and 0x45). Thus, FPC would only store the first byte from each half-word, in addition to the prefix (0b101). On the other hand, the last word is zero, so FPC will store the prefix (0b000) and the three-bit run

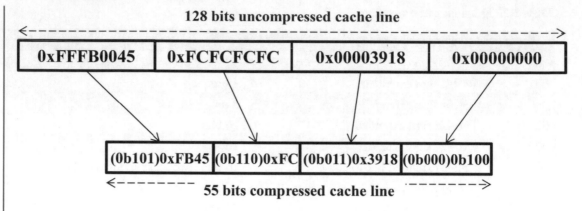

Figure 2.4: An example of FPC compression algorithm.

length (value 0b100), in this example). In this way, FPC will store the 128-bit uncompressed block as 55-bit compressed format.

2.3.6   BASE-DELTA-IMMEDIATE (BDI)

Base-Delta-Immediate (BDI) compression [19] is a low-overhead general-purpose algorithm for compressing data in on-chip caches and memory. It is based on exploiting spatial value locality, i.e., the observation that values that are spatially close in memory also tend to have small differences in their values. BDI represents a block using one or more base values and an array of differences from the base values. Figure 2.5 illustrates one example of BDI compression using a single base value. BDI is essentially a generalization of significance-based compression, using one or more explicit bases instead of an implied base of zero. However, determining the optimum set of base values is complicated. Instead, a typical implementation of BDI uses two base values: zero and the first non-zero value in the input message. This simple selection scheme allows BDI to compress/decompress all words in the block in parallel.

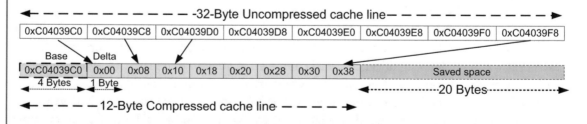

Figure 2.5: BDI compression using one Base value [19].

2.3.7 CACHE PACKER (C-PACK)

C-PACK [18] is a compression algorithm designed specifically for hardware-based cache compression. C-PACK combines a significance-based algorithm with a dictionary, thus it is both a temporal-value and spatial-value-based algorithm. C-PACK treats data blocks as a stream of 4-byte words and detects and compresses frequently appearing words (such as zero words) to fewer bits. In addition, it also uses a small dictionary to compress repeated input symbols, including partial matches. Table 2.3 shows how C-PACK encodes different patterns, with "b' and "B' representing one and four bits in the output codeword, respectively. For a 64-byte block, the dictionary has 16 entries, each storing a 4-byte word. The dictionary is built and updated on the fly for each data block. Before compressing a cache block, C-PACK initializes the dictionary to zeros. C-PACK checks whether each input symbol matches a dictionary entry (even partially), or matches any of the static patterns in Table 2.3. Figure 2.6 shows a simple example for C-PACK compressing four input words (thus only four directory entries are shown). The first input symbol "0x12345678" naturally does not match a dictionary entry, nor does it match any of C-PACK's significance-based patterns. Thus, it is added to the dictionary and output as an uncompressed word with the prefix "0b01" (represented with the notation "(01)"). The next input, "0x123456AA," matches the three most-significant bytes of the first dictionary entry. Thus, C-PACK generates the three-byte-match prefix code "0b1110" and dictionary index "0b0000" as the code word, and also updates the dictionary. The third input, "0x000000AB," does not match a dictionary entry, but does match the one-byte zero-extended pattern. Thus, C-PACK outputs the prefix code "0b1100" and the unique byte ("0xAB"), and also inserts the word in the dictionary. Finally, the last input "0x12345678" exactly matches the first dictionary entry, resulting in the prefix code "0b10" and the dictionary index "0b0000" being output. During decompression, C-PACK again initializes the dictionary, then builds it back up from stream of code words. To make compression/decompression faster, C-PACK can process multiple words in parallel. However, processing multiple words in parallel requires a parallel prefix circuit, due to the sequential dependence on the dictionary (i.e., an insert of word X should be found by word X+1). To limit

Table 2.3: C-PACK pattern encoding

Prefix	Pattern	Output Codeword	Length (bits)
00	zzzz (zero)	(00)	2
01	xxxx (uncompressed)	(01)BBBB	34
10	mmmm (matched a dictionary entry)	(10)bbbb	6
1100	mmxx (matched 2 bytes with a dictionary entry)	(1100)bbbbBB	24
1101	zzzx (one byte zero-extended)	(1100)B	12
1110	mmmx (matched 3 bytes with a dictionary entry)	(1110)bbbbB	16

	After Input 0		After Input 1		After Input 2		After Input 3	
Dictionary	0000	0x12345678	0000	0x12345678	0000	0x12345678	0000	0x12345678
	0001		0001		0001		0001	
	0010		0010		0010		0010	
	0011		0011		0011		0011	
Input Symbol	0x12345678		0x123456AA		0x000000AB		0x12345678	
Action	Emit uncompressed Update dictionary		Emit 3-byte match Update dictionary		One-byte sign extend Update dictionary		Full match	
Code Word	(01) 0x12345678		(1110)(0000)0xAA		(1100)0xAB		(10)(0000)	
Code Word Size	34 bits		16 bits		12 bits		6 bits	

Figure 2.6: An example for C-PACK.

the circuit complexity and meet timing, the C-PACK implementation compresses/decompresses only two words on each cycle.

2.3.8 DEDUPLICATION

While most compression algorithms focus on compressing a single input message or cache block, deduplication seeks to identify and exploit redundancy across multiple memory blocks. Deduplication has been widely used in virtual memory systems and virtual machines, where null pages and shared pages (e.g., of shared libraries) can result in significant redundancy across multiple processes. Virtual address translation hardware facilitates deduplication, allowing the operating system to map multiple virtual addresses to point to the same physical memory page (i.e., creating virtual address synonyms). Implementing a copy-on-write policy allows the operating system to de-deduplicate pages when necessary.

Deduplication has also been explored in the context of instruction caches [119], last-level caches [115, 118], and main memory [118, 124]. Implementing deduplication in hardware is challenging for several reasons. First, a memory access has to be redirected from one location to another. Second, on an update, a block that was deduplicated must be disassociated from the other deduplicated blocks. More importantly, detection of duplication is an expensive process, therefore hashing is typically used to reduce the cost of comparisons.

In CATCH [119], Kleanthous and Sazeides propose a mechanism to detect duplicated cache blocks in instruction caches by correlating addresses that contain the same value, a technique that is known as address correlation [125]. CATCH is composed of different structures. First, there is a Duplication Relation table (DR) that stores the correlations between two PC addresses. On a cache miss, the DR is accessed: If it is a DR hit, the second PC of the entry is read to locate the duplicate block in the cache; otherwise (DR miss), the block is fetched from the lower-level cache. In the case of the DR miss, the Hashed-Duplicate-Detection (HDD) cache structure is used. The HDD contains hash-codes (one hash-code per entry), which are generated based on the block content. If the currently fetched block matches an HDD hash code, then the two blocks are further compared by the Block Compare Unit (BCU). On a match, an entry is created in the DR table, while in case of mismatch (i.e., false positive), the block is inserted into the cache and the HDD entry is updated to point to the associated PC.

Tian et al. [115] propose a deduplication approach also at the block level but for last-level caches. Duplication is detected by using an augmented hash table which is implemented as a sequence of buckets. Hashing is implemented using XOR gates organized in five levels. Each node in the bucket contains a pointer to the data array. The proposed deduplication mechanism changes the organization of the cache by decoupling the data array from the tag array so that multiple tags can be associated with one distinct block. This is similar to prior compression approaches, as is discussed later in Chapter 3, but the two arrays operate as two different structures. The tags keep all the status information (dirty, replacement, etc.). The tags that are correlated with one data block are organized in doubly linked lists, thus two pointers are required per tag as well as an extra tag-to-data pointer. On the other hand, the data array contains a counter and a data-to-tag pointer (list's head) per cache line. The counter is used to keep track of the number of tags associated with this line but also drives the replacement. The authors find that each block has 2.23 duplicates, on average, in a last-level cache when SPEC2006 program mixes run in a 4-core processor.

Cheriton et al. [118] propose Hicamp as a deduplication solution for cache and main memory. Hicamp performs deduplication by studying finer granularities than prior work: block sizes between 16 bytes and 64 bytes. Duplicate data blocks are correlated and tracked by implementing hashing using a Content Addressable Memory (CAM) scheme in DRAM and storing one hash bucket per DRAM row. The authors see most of the benefit for 16-bytes lines, however the finer the granularity the more metadata is needed. Another interesting characteristic of the Hicamp solution is that a programming model interacts with the deduplication mechanism and provides tips about candidate blocks for deduplication targets, accelerating this way detection of duplicated blocks. Finally, Biswas et al. [124] propose a novel memory allocation library (SBLL-malloc), which merges identical memory blocks in the same copy at allocation time reporting a memory footprint reduction by 30%, on average.

2.3.9 INSTRUCTION COMPRESSION

General-purpose compression algorithms usually perform poorly for instruction blocks as instructions have more complicated bit patterns than many data blocks, due to the presence of multiple tightly packed encoded fields. Several specialized compression techniques have been proposed to improve compressibility of instructions. Instruction compression is in particular important in embedded systems, where instruction storage may be more expensive. Instruction compression can also improve performance by effectively increasing instruction fetch bandwidth.

Most instruction compression techniques find frequently used instruction sequences in the instruction stream, replacing those with small code words to reduce instruction size [1–10]. Thus, these are temporal-value-based algorithms. For example, Lefurgy et al. [1] propose a post-compilation analyzer that examines a program, and replaces common instruction sequences with small code words. The processor fetches these code words and expands them to the original sequence of instructions in the decode stage. Their technique benefits programs in embedded processors where instruction memory size is expensive. Benini et al. [5] similarly compress the most commonly executed instructions to reduce energy in embedded systems. They decompress instructions on the fly by a hardware module located between the processor and memory.

Cooper et al. [6] explore compiler techniques for reducing memory needed to load and run program executables for an RISC-like architecture. They reduce instruction size using pattern-matching techniques to identify and coalesce together repeated instruction sequences. Similarly, Wolfe and Chanin [7] target reducing the instruction size of RISC architectures using compression. They designed a new RISC system that can directly execute compressed programs. They use an instruction cache to manage compressed programs. The processor executes instructions from the cache, so the compression is transparent to the processor.

Thuresson and Stenström [10] evaluate the effectiveness of different dictionary-based instruction compression techniques in reducing instruction size. Dictionary-based instruction compression techniques statically identify identical instruction sequences in the instruction stream and replace them by a code word. Later, at runtime, they replace the code word by the corresponding instruction sequence (i.e., the dictionary entry). The authors show that this technique can reduce instruction size significantly.

Thuresson et al. [11] address increased instruction-fetch bandwidth and larger instruction footprint in VLIW systems using compression. They compress at compile time by analyzing which subset of a wide instruction set is used in each basic block based on profiling. They also propose a decompression engine that comprises a set of tables that dynamically convert a narrow instruction into a wide instruction.

2.3.10 FLOATING-POINT COMPRESSION

Similar to instructions, floating-point data is generally not compressible with general-purpose compression algorithms. Floating-point numbers are usually represented in binary with three binary fields that carry semantic information: sign, exponent, and mantissa. For example, single-precision 32-bit IEEE floating-point numbers use a 1-bit sign, which shows whether the number is positive (sign = 0) or negative (sign = 1), an 8-bit exponent with an added bias of 127, and a

23-bit mantissa representing 223 uniformly spaced numbers within the range associated with a particular exponent. This representation makes compressing floating-point numbers hard.

There are several proposals to improve compression for floating-point data, most of which seek to exploit spatial-value locality. Isenburg et al. [12, 13] propose a compression technique to reduce storage size for floating-point geometric coordinates in scientific and industrial applications. They propose a lossless compression technique using predictive coding. For each coordinate, they predict values in floating-point and compressed the corrections from the actual value using context-based arithmetic coding. Lindstrom and Isenburg [14] also present an online lossless compression of floating-point data to accelerate I/O throughput in real simulation runs. They also use prediction, and for each data value, they predict it from previously encoded data. They then compressed the difference between the actual and predicted value.

Ratanaworabhan et al. [15] propose the FPC algorithm to compress sequences of IEEE double-precision floating-point values (note that despite having the same acronym, this algorithm is quite different from Frequent Pattern Compression discussed in Section 2.3.5). They use value prediction, predicting each value in the sequence, and XORed it with the true value; the true value is the outcome of right shifting the initial binary floating-point value by 48 bits, thus keeping the sign bit, the exponent and the four most significant bits of the mantissa. They then encode and compress the residual by dropping the high-order zero bits (leading-zero compression). In another work [16], the authors further extend FPC to double-precision floating-point data.

In contrast, Arelakis et al. propose FP-H [127], a temporal-value-based algorithm for floating-point data compression. FP-H compresses the different semantic fields of floating-point values in isolation using Huffman encoding. Moreover, the authors observe that the mantissa exhibits value locality if it is further partitioned, e.g., its 20 most significant bits when the mantissa is partitioned into two sub-fields, however other ways of partitioning may work for different workloads.

2.3.11 HYBRID COMPRESSION

As described in the previous sections, there are many special-purpose compression techniques or methods that target specific data types and semantics. Based on the type, they make design time assumptions on an approach to exploit the respective value locality sub-property exhibited by a data type. Today's computer systems process on various data types at the same time; hence, data of diverse types are stored in the memory hierarchy or transmitted via the on/off-chip network. Because of this type diversity, none of the special-purpose compression methods/techniques is found always better than others. What is needed is a technique that selects the compression method that will perform the best given the target type/semantics. However, the underlying hardware is unaware of the implicit data types of the binary data transferred or stored in the cache/memory.

HyComp [127] proposes a hybrid compression method tailored for caches that uses heuristics to predict the best (in terms of compression ratio) performing compression method to compress a cache block. HyComp makes an association between data types and specific compression methods. The target data types are integers, floating-point, and pointers and are associated with SC^2 [46] (Section 2.3.2), FP-H [127] (Section 2.3.9), and BDI [19] (Section 2.3.5), respectively. Null blocks, despite not being a data type, are common and are included by HyComp to be compressed with ZCA [37]. As illustrated in Figure 2.7 every block that is inserted into the LLC from the main memory or written back from the upper-level caches is scanned in the HyComp unit that implements the heuristics. Based on the heuristics' outcome, a compression method is selected to compress the block before being stored in the LLC and the selection is recorded in the tag so that when the block is requested from the upper-level caches or evicted to the main memory, it is known which decompressor must be used.

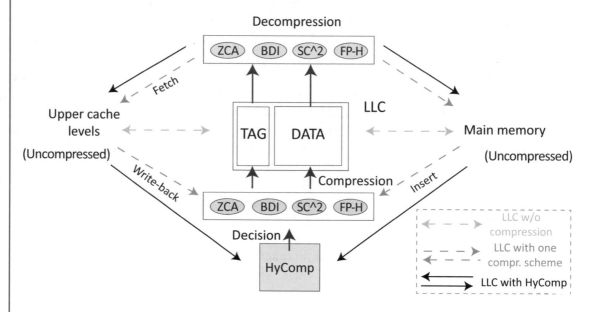

Figure 2.7: HyComp (Hybrid Compression) [127].

2.4   METRICS TO EVALUATE THE SUCCESS OF A COMPRESSION ALGORITHM

There are several metrics that can be used to evaluate the success of a compression algorithm, including compression ratio, compression and decompression latency, and the area and power overheads of compression and decompression units. The compression ratio (also referred to as the compression factor) is defined as the size of the original uncompressed data divided by the size of the compressed code words. The higher the compression ratio, the higher the potential compression benefits (e.g., saved space or bandwidth). However, there are usually trade-offs between the compression ratio and the overheads of compression/decompression. The more complex algorithms may result in higher compression ratios. However, they usually result in higher overheads, including compression and decompression latency, area, and power. Thus, many hardware compression approaches have favored simple algorithms with lower overheads, but also with lower compression ratios [19, 20].

How best to resolve these trade-offs depends on the design point. For example, in compressed caches and memory, decompression latency tends to be particularly important as it lies on the critical memory access path and can degrade performance. This is especially true for L1 and L2 caches, since performance is highly sensitive to their latency. But as multicore systems move to having three or more levels of cache, the sensitivity to the lower (larger) level cache latencies decreases, allowing systems to consider more effective, longer latency compression algorithms. In addition, many systems use different mechanisms to hide memory latency, such as out-of-order (OOO) or multi-threaded cores. Those systems are better able to tolerate higher decompression latency, so they could possibly benefit from more complex algorithms. Similarly, more complex algorithms are better suited to the main memory than caches, where the cache hierarchy can effectively hide the extra latency for many workloads. Arelakis and Stenström [46] recently showed how an aggressive compression algorithm like Huffman coding can be suitable for last-level caches. Earlier, IBM MXT decided to use a complex (and relatively slow) compression algorithm to increase memory capacity [30].

2.5 SUMMARY

This chapter focuses on state of the art compression algorithms. The chapter first introduces the value locality property, which is another property that is exhibited by data in addition to the property of reference locality. Based on this, it makes a first categorization of the compression algorithms depending on the value locality sub-property they exploit. It then presents a taxonomy of the compression algorithms with respect to three criteria: I) general purpose vs. special purpose, II) static vs. dynamic, and III) temporal-value based vs. spatial-value based. Based on the taxonomy, it studies the most popular algorithms that have been explored for cache/memory and link compression. Finally, the chapter concludes with a discussion about the metrics used to evaluate the efficiency of those compression algorithms.

CHAPTER 3

Cache Compression

Caches have long been a critical component of the memory hierarchy. Early work focused on using caches to reduce effective access time or latency. Later, caches were also used to reduce required memory bandwidth. In modern multicore processor systems, where the memory hierarchy accounts for a large fraction of total system energy, caches play a critical role in reducing energy as well. Caches filter out expensive off-chip memory accesses, and replace them with much cheaper cache accesses [129]. Increasing cache size can improve system performance and energy by reducing memory accesses, but at the cost of high area and power overheads at the cache level.

Cache compression has the potential to expand the effective cache capacity with little area overhead. Compressed caches can achieve the benefits of larger caches using the area and power of smaller caches by fitting more cache blocks in the same cache space. Or, equivalently, compression can allow a cache to use much less area to hold a comparable amount of data. Designing a compressed cache typically has two main parts: a compression algorithm to represent the same data blocks with fewer bits, and a compaction mechanism to fit compressed blocks in the cache. In Chapter 2, we presented several compression algorithms that exploit redundancy to compress data. For hardware cache compression, the latency, area, and power overheads of the compression and decompression algorithms matter, in addition to their ability to achieve a good compression ratio.

For a given compression algorithm, the cache compaction mechanism—how to store and track more compressed cache blocks in the same space—is important. In order to track more blocks in the cache, a compressed cache needs extra tags and metadata. An ideal design would fit variable size compressed blocks tightly to reduce internal fragmentation, while keeping tag and metadata overheads low. In practice, however, there are several trade-offs to consider.

In the rest of this chapter, we first present a taxonomy of cache compaction mechanisms in Section 3.1, then in Section 3.2 we explain how previously proposed cache compaction schemes work. We then present several policies proposed to improve efficiency of compressed caches in Section 3.3, and discuss techniques that mainly employ cache compression to improve cache area and power in Section 3.4.

3.1   CACHE COMPACTION TAXONOMY

A compression algorithm maps a fixed size data message (e.g., cache block) to variable size code words at bit or byte granularity, while conventional caches operate on fixed-size blocks (e.g., 64 bytes). Thus, to achieve the potentials of a given compression algorithm, the compaction mechanism plays a critical role to manage compressed blocks in the cache. Table 3.1 introduces a taxonomy of the current state of the art. We classify previous work using three main design factors: (1) *how to provide the additional tags and metadata*, (2) *allocation granularity of compressed blocks*, and (3) *how to find the corresponding block given a matching tag*.

Table 3.1: Compressed caches taxonomy

Tags			Data		
			Half-Block	Sub-Block	Byte
Per Block	Direct One-to-One Tag Mapping		CC[39] Lee et al. [47][48] Significance-compression [45]	—	—
	Decoupled Forward Pointers		—	VSC [20]	SC2[46]
	Decoupled Back Pointers		—	IIC-C [50]	—
Per Super Block	Direct One-to-One Tag Mapping		—	SCC [23]	—
	Decoupled Forward Pointers		—	—	—
	Decoupled Back Pointers		—	DCC [21][22]	—

Number of Tags: Conventional caches store one address tag for each fixed-size data block. To track more blocks, compressed caches require additional tags for the same amount of data storage. Many compressed cache designs simply double the number of tags (i.e., 2x Block Tags), allowing them to track up to twice as many cache blocks in the cache [20]. Naively increasing the number of tags increases the cache tag overhead, which is particularly costly for LLCs, since they are already one of the largest on-chip components. Decoupled Compressed Cache (DCC) [21, 22] and Skewed Compressed Cache (SCC) [23] use an alternative approach to increase the number of tags with low area overhead. DCC and SCC exploit spatial locality and use super-block tags to effectively track more blocks while keeping the overheads low. DCC, for example, uses the same number of tags as a regular cache, but each tag tracks a 4-block super-block (i.e., using Super-Block Tags instead of regular block tags), and can map up to four cache blocks. Tracking super-blocks only slightly increases tag area compared to the same size regular cache.

Allocation Granularity: A given compression algorithm will (usually) convert a fixed-size cache block into a variable, but smaller, number of bits representing the same information. Like all memory allocators that manage variable-size data types, there are trade-offs between internal and external fragmentation. At one extreme, allocating space at the bit granularity eliminates internal fragmentation, but incurs high overheads to manage such fine-grain allocations and makes it difficult to reuse the highly variable sizes that may result. Conversely, at the other extreme— allocating a full cache block regardless of the data's compressibility—adds no additional management complexity, but results in high internal fragmentation and eliminates any increase in capacity from cache compression. Previous cache compaction mechanisms have explored allocating data at various granularities—half-blocks, 16-byte, or 8-byte sub-blocks, or bytes—trading off internal fragmentation for simpler management of the data space.

Tag-Data Mapping: The final issue is how the cache maintains the mapping between address tags (and associated metadata) and the data. Traditional caches maintain a *direct one-to-one relationship* between tags and data, so a matching tag implicitly identifies the corresponding data. Compressed caches seek to pack more blocks into the same data space, and thus their address tag arrays must be able to map more blocks than a conventional cache of the same data capacity. This additional tag "reach" can be provided using more per-block tags (e.g., twice as many tags), using superblock tags (e.g., that map two or four contiguous blocks), or some combination. Compressed caches usually require a many-to-many mapping between tags and data, requiring some form of decoupled tag organization. There is also an interaction between the allocation granularity and the tag-data which affects the amount of metadata needed to track compressed blocks.

3.2   CACHE COMPACTION MECHANISMS

3.2.1   SIMPLE COMPACTION MECHANISMS

The earliest compressed caches maintain a direct tag-data relationship by allowing only one compressed size (i.e., half the block size). Yang et al. [39] exploited the value locality phenomenon to design a first-level compressed cache (Compression Cache). Each cache line of the Compression Cache (CC) stores either one uncompressed line or two lines compressed to less than half their original sizes [39]. Figure 3.1 illustrates the tag-data mapping for different cache designs. In a regular cache, there is a one-to-one correspondence between tags and data entries. For example, Figure 3.1(1) illustrates one set of an uncompressed two-way set-associative cache, which has two tags per set, each with a corresponding data entry. Figure 3.1(2) illustrates a simple compressed cache, like CC [39], that doubles the number of tags per set. They divide each data entry into half, so that each tag corresponds to half a data entry. A block is compressed to half a block if possible, such as blocks A and B in Figure 3.1(2). Otherwise, the block would be stored as uncompressed, such as block C in Figure 3.1(2). In this way, the cache can store up to two times more blocks (if all compressible to half).

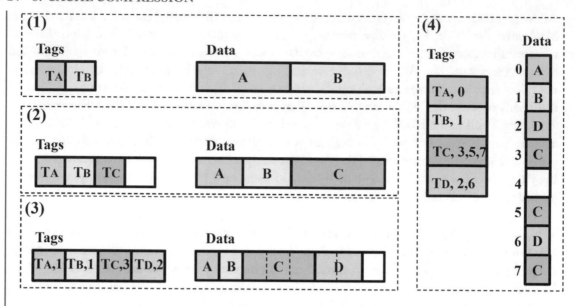

Figure 3.1: Alternative tag-data mappings in regular and compressed caches.

Lee et al. [47, 48] proposed a compressed cache that selectively compresses cache blocks if and only if they can be compressed to less than half their original size. That compressed cache reduces the decompression overhead [47] by selectively compressing blocks only if their compression ratio is less than a certain threshold, decompressing blocks in parallel, and buffering recently accessed blocks at the L2 cache in an uncompressed format. In another work, Lee et al. [48] propose a cache design that compacts block pairs, storing them in the space of one uncompressed block if both blocks can be compressed by 50% or more. This frees a cache block in an adjacent set; however, its needs checking two sets for a potential hit on every access, which increases power overheads. Kim et al. [45] also compressed cache blocks into half using a significance-based compression scheme to improve cache utilization. Their proposal compressed and stored a cache block as a half block if the block's upper half was either all zeros or all ones. Otherwise, the cache stored the whole block as uncompressed. Overall, these techniques limit compressibility by failing to take advantage of blocks that compress by less than 50% (e.g., block C in in Figure 3.1(2)), and so introducing internal fragmentation.

3.2.2   SUPPORTING VARIABLE SIZE COMPRESSION

To reduce internal fragmentation, Alameldeen and Wood [20] presented a compressed cache that decouples tag-data mapping [20]. Their proposal, which we refer to as Variable Size Compression (VSC), compacts compressed blocks into a variable number of 8-byte sub-blocks (called segments in the original work), using the FPC compression algorithm [20]. Figure 3.1(3) shows a simple example of VSC for a two-way associative cache. VSC reduces internal fragmentation, since all blocks in a set share the same pool of sub-blocks. It stores a compressed block into contiguous sub-blocks in its corresponding data set. For example, blocks A and B in Figure 3.1(3) are compressed to one sub-block each, while blocks C and D are compressible to three and two sub-blocks, respectively.

VSC decouples the tag-data mapping and thus needs additional metadata to track which tag is associated with which data. VSC does this by keeping a block's compressed size as additional metadata with each tag. For example, in Figure 3.1(3), the first tag indicates that block A has only one sub-block. On a lookup, VSC finds the block by adding up the size of all previous blocks in its corresponding set. For example, the data for block D in Figure 3.1(3) will be stored starting at sub-block $5(1 + 2 + 3 = 5$ sub-blocks). Using this technique, VSC keeps the metadata overhead low. However, on a block update, when the block's compressibility and size might change, VSC may require moving other blocks in the corresponding set in order to make enough contiguous space for the accessed block. This is called re-compaction. For example, consider an update to block A in Figure 3.1(3) that changes its compressed size from one to two sub-blocks. VSC will need to move blocks B, C, and D (requiring one extra read and one extra write) to make two free contiguous sub-blocks for the now larger block A. As updates can happen frequently, re-compaction may incur high dynamic energy overheads.

To increase cache utilization, SC^2 [46] also decouples tags from data, but compresses blocks into a variable number of bytes using a variable-length Huffman encoding. Each compressed block can be stored at any byte position the cache set's region of the data array. The starting byte of a compressed block is located by storing a "forward" pointer as metadata with each tag. For example, for a 16-way set-associative cache with 64-byte blocks, SC^2 stores 10 extra bits per tag to locate the block in its corresponding data set, $(\log_2(16 \times 64) = 10)$. As a compressed block is allocated into contiguous sub-blocks, SC^2 defines a set of policies to manage the free space, while it modifies the LRU replacement policy to enforce eviction of more than one block. If the new or modified block is larger than the old, adjacent blocks are evicted to create space. On the other hand, if the new/updated block is smaller, the released space is allocated to the adjacent blocks if they are invalid, or block compaction is performed to create contiguous space in the LRU block.

Hallnor and Reinhardt [50] extend their earlier indirect index cache [49] to support compression, a design they call IIC-C. IIC-C compresses blocks into a variable number of sub-blocks using the LZSS algorithm [32]. Unlike VSC and SC^2, IIC-C eliminates re-compaction overhead by allocating the sub-blocks of a block anywhere in the data array (not just in the region corresponding to the tag's set). However, this flexibility requires each tag to have a forward pointer per sub-block and each forward pointer must be large enough to point to any sub-block in the data array. Figure 3.1(4) shows a simple example of ICC-C. Block C, for example, is compressed and stored in three non-contiguous sub-blocks in the data array (i.e., blocks 3, 5, and 7). The corresponding tag entry maintains pointers to each of these sub-blocks. The overhead of the forward pointer metadata in ICC-C can be very large. For example, in an 8MB LLC with 64-byte blocks,

16-byte sub-blocks, and twice the original number of tags, IIC-C incurs about 24% area overhead or the equivalent of approximately 2MB of cache. Further increasing the number of tags increases the area overhead, but provides an opportunity for higher effective cache capacity.

3.2.3   DECOUPLED COMPRESSED CACHES

Sardashti and Wood propose Decoupled Compressed Caches (DCC) [21, 22]. Like VSC, DCC uses sub-blocks to balance internal fragmentation and metadata overheads. DCC uses decoupled super-block tags and "backward" pointers to reduce tag and metadata overheads. Figure 3.2(a) shows the key components of DCC for a two-way-set associative cache with four-block super-blocks, 64-byte blocks, and 16-byte sub-blocks. DCC has three main data structures: a Tag Array, a Sub-Blocked Back Pointer Array, and a Sub-Blocked Data Array. All three structures are indexed using the super-block address bits (Set Index in Figure 3.2(e)). Unlike conventional super-block caches, this means that all blocks of the same super-block are mapped to the same data set.

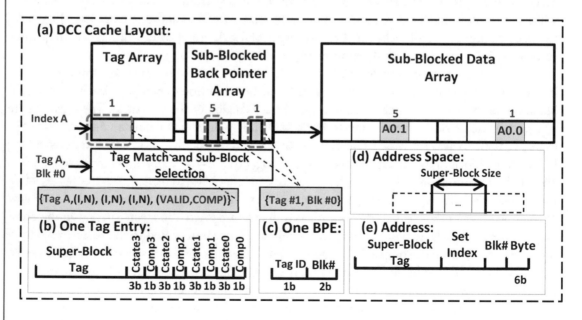

Figure 3.2: DCC cache design.

DCC explicitly tracks super-blocks through the tag array. The tag array is a largely conventional super-block tag array. Figure 3.2(b) shows one tag entry that consists of one tag per super-block (Super-block tag) and coherence state (CState) and compression status (Comp) for each block of the super-block. Since all four blocks of a super-block share a tag address, the tag array can map four times as many blocks as the same size conventional cache with minimal area overhead. DCC holds as many super-block tags as the maximum number of uncompressed blocks that could be stored. For example, in Figure 3.2, for a two-way-associative cache, it holds two super-block tags in each set of the tag array. In this way, each set in the tag array can map eight blocks (i.e., 2 super-blocks * 4 blocks/super-blocks), while a maximum of two uncompressed blocks can fit in each set. In the worst-case scenario, when each super-block is a singleton (i.e., contains only one valid block) or the data is uncompressible, DCC can still utilize all the cache data space, for example, by tracking two singletons per set.

DCC compacts cache blocks into a variable number of (potentially) non-contiguous sub-blocks. The data array is a mostly conventional cache data array, with each set organized in sub-blocks. Figure 3.2(a) illustrates a cache with eight 16-byte sub-blocks per set, for a total of 128 bytes. This is only one quarter of the data space mapped by each set in the tag array (i.e., 2 super-blocks * 4 blocks/super-block * 64 bytes/block = 512). Thus, using this configuration the tag array has the potential to map four times as many blocks as can fit in the same size uncompressed data array.

DCC decouples sub-blocks from the address tag to eliminate expensive re-compaction when a block's size changes. Unlike VSC [20], which requires contiguous sub-blocks for a block, DCC allows the sub-blocks of a block to be non-contiguous. For example, in Figure 3.2(a), block A0 is compressed into two sub-blocks (A0.1 and A0.0) that are stored in the sub-block #5 and the sub-block #1 in the data array.

DCC uses "back" pointers to maintain the decoupled tag-data mapping while keeping the overhead low. Back pointers are logically associated with each sub-block and point back to the corresponding address tag. For each sub-block in the data array, the back pointer array keeps one back pointer entry (BPE) for each sub-block. A BPE encodes the owner block of a sub-block by storing the matched tag ID (i.e., the super-block tag entry tracking the owner block), and the corresponding Block ID (i.e., the block number in its corresponding super-block). In Figure 3.2, for example, for a two-way associative cache with four-block super-blocks, DCC stores eight BPEs (2 * 4 sub-blocks per set), each of three bits (1-bit tag ID + 2-bit block ID), for a total of 24 bits. Like IIC-C, DCC eliminates re-compaction when a block's size changes, because sub-blocks need not be allocated contiguously. But DCC requires far less metadata overhead for two reasons. First, DCC restricts tags and data to be stored in the same set, thus pointers require only a few bits. Second, using back pointers requires one fourth the metadata overhead of forward pointers, which would require 32 forward pointers (2 tags per set * 4 blocks per tag * 4 sub-blocks per block * 1 pointer per sub-block), each of which is three bits (points to one of eight sub-blocks in the data array), for a total of 96 bits.

3.2.4   SKEWED COMPRESSED CACHES

Sardashti et al. propose another compressed cache called Skewed Compressed Cache (SCC) [23] that allocates variable size compressed block while eliminating the need for extra metadata to track blocks. Similar to DCC, SCC compacts blocks into a variable number of sub-blocks to reduce internal fragmentation, but retains direct tag-data mapping to find blocks quickly and eliminate extra metadata (i.e., no forward or back pointers).

SCC builds on the observation that most workloads exhibit (1) spatial locality, i.e., neighboring blocks tend to simultaneously reside in the cache; and (2) compression locality, i.e., neighboring blocks often have similar compressibility [61]. SCC exploits both types of locality to compact neighboring blocks with similar compressibility in one physical data entry (i.e., 64 bytes) if possible. Otherwise, it stores neighbors separately.

Figure 3.3 illustrates SCC functionality using some examples. This figure shows a 16-way cache with eight cache sets. The 16 cache ways are divided into four way groups, each including four cache ways. For the sake of clarity, Figure 3.3 only illustrates super-blocks that are stored in the first way of each way group. This example assumes 64-byte cache blocks, eight-block super-blocks, and eight-byte sub-blocks, but other configurations are possible. A 64-byte cache block can compress to any power-of-two number of eight-byte sub-blocks (i.e., one, two, four, or eight sub-blocks). Eight aligned neighbors form an eight-block super-block. For example, blocks I—P belong to SB2.

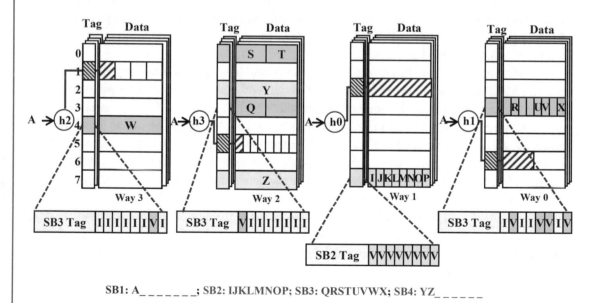

Figure 3.3: Skewed compressed cache.

SCC differs from a conventional cache by storing a sparse super-block tag per data entry. Like a conventional super-block cache, SCC's tags provide additional metadata that can track the state of a group of neighboring blocks (e.g., up to eight aligned, adjacent blocks). However, SCC's tags are sparse because—based on the compressibility of the blocks—they may map only one (uncompressed), two, four, or eight compressed blocks. This allows SCC to maintain a conventional one-to-one relationship between a tag and its corresponding data entry (e.g., 64 bytes).

SCC only maps neighboring blocks with similar compressibility to the same data entry. For example, if two aligned, adjacent blocks are each compressible to half their original size, SCC will allocate them in one data entry. This allows a block's offset within a data entry to be directly determined using the appropriate address bits. This eliminates the need for additional metadata (e.g., back pointers in DCC [21]) to locate a block. Figure 3.3 illustrates different examples. If all eight neighbors exist and are compressible to one eight-byte sub-block each, SCC will compact them in one data entry, tracking them with one tag. For example, in Figure 3.3, all blocks of SB2 are compacted in one data entry in set #7 of way #1. SCC tracks them with the corresponding tag entry with the states of all blocks set as valid (V in Figure 3.3). If all cache blocks were similarly compressible, SCC would be able to fit eight times more blocks in the cache compared to a conventional uncompressed cache. On the other hand, in the worst-case scenario when there are only singletons (i.e., only one out of eight neighbors exists in the cache) or blocks are not compressible, SCC can still utilize all cache space by allocating each block separately. For example, in Figure 3.3, only blocks Y and Z from SB4 are present in the cache, and neither are compressible. Thus, SCC stores them separately in two different sets in the same way group, tracking them separately with their corresponding tags.

SCC's cache lookup function is made complex because the amount of data mapped by a sparse super-block tag depends upon the blocks' compressibility. SCC handles this by using a block's compressibility and a few address bits to determine in which cache way(s) to place the block. For example, for a given super-block, uncompressed blocks might map to cache way #0, blocks compressed to half size might map to cache way #2, etc. For instance, in Figure 3.3, block A maps to a different set in each cache way depending on its compressibility, shown in hatched (red) entries. SCC allocates A in way group #0, #1, #2, or #3 if A is compressible to 32 bytes (four sub-blocks), 64 bytes (eight sub-blocks), eight bytes (one sub-block), or 16 bytes (two sub-blocks), respectively. These mappings would change for a different address, so that each cache way would have a mix of blocks with different compression ratios. This is important, as it permits the entire cache to be utilized even if all blocks compress to the same size. For instance, SCC allocates block A and block I in cache way #1, if A is uncompressible and I is compressed to eight bytes (one sub-block). Using this mapping technique, for a given block, its location determines its compression ratio. This eliminates the need for extra metadata to record block compressibility.

To prevent conflicts between blocks in the same super-block, SCC uses different hash functions to access ways holding different size compressed blocks. On a cache lookup, the same address bits determine which hash function should be used for each cache way. Like all skewed associative caches, SCC tends to have fewer conflicts than a conventional set-associative cache with the same number of ways. SCC maps blocks to different cache ways based on their compressibility, using different index hash functions for each cache way [92]. To spread all the different compressed sizes across all the cache ways, the hash function used to index a given way is a function of the block address. SCC skews compressed blocks across sets within a cache way to decrease conflicts [93, 94] and increase effective cache capacity. Compared to DCC, SCC achieves comparable or better performance, with a factor of four lower area overhead, a simpler data access path, and a simpler replacement policy.

3.3   POLICIES TO MANAGE COMPRESSED CACHES

Adaptive compression: Compressed caches introduce a trade-off between cache capacity and cache access latency. On the one hand, they can improve effective cache capacity by storing more cache blocks, resulting in a possibly lower cache miss rate. On the other hand, they incur higher access latency as they decompress compressed blocks. These trade-offs change depending on different parameters, including the sensitivity of the applications to cache latency and capacity, cache level and decompression latency. For capacity-sensitive workloads, compression can improve performance by reducing costly misses to the next level of hierarchy, while for cache-insensitive workloads or latency-sensitive workloads, the latency overhead of decompression can impact performance. The overhead is higher with longer latency decompression techniques and at lower levels of cache hierarchy (L1 or L2).

To balance this trade-off, Alameldeen and Wood [20] propose an adaptive policy that dynamically adapts to the costs and benefits of cache compression. They consider the benefit of cache compression to be the number of *Avoided Misses ∗ Miss Penalty* and the (performance) cost to be the *Penalized Hits ∗ Decompression Latency*. If the benefits exceed the costs, then the cache should compress blocks, and not otherwise.

They considered a two-level cache hierarchy, with uncompressed L1 caches and adaptively compressed L2 cache. On a cache allocation, the L2 cache would compress a new block if the past benefits exceed the past costs. To determine whether a reference is an avoided miss or a penalized hit, the cache uses the LRU stack depth. If the LRU stack depth is greater than the (uncompressed) associativity, then compression avoided a miss. If the LRU stack depth is less than the (uncompressed) associativity and the block was compressed, then the access is a penalized hit. To efficiently track the cost-benefit trade-off, the cache keeps a global saturating counter. On each cache access, it increments this counter by the L2 miss penalty if compression could avoid a miss, and decrements the counter by the decompression latency if the access would have been a hit even without compression. Using this counter, the cache predicts whether to allocate future cache lines in compressed or uncompressed form. By dynamically monitoring workload behavior, this adaptive compressed cache achieves the benefits of compression for cache sensitive workloads, while avoiding performance degradation for others. The adaptive mechanism can be used in other compressed caches and at other levels of cache hierarchy as long as we are using an LRU replacement policy.

Tailored replacement policy: Compressed caches typically use the same replacement policy as traditional caches that treat all blocks similarly, while the sizes of the cache blocks vary depending on their compressibility. Baek et al. [55] propose a size-aware compressed cache management, Effective Capacity Maximizer (ECM), to improve the performance of compressed caches. ECM uses cache block size as a hint to select a victim to improve cache performance. In a compressed cache, the eviction overhead varies based on the size of the evicted and incoming cache blocks. If the size of the new block is larger than the victim block, the compressed cache needs to evict more blocks. Thus, ECM considers block size in the cache management policies to increase effective capacity. It classifies blocks as big-size or small-size based on their compressed size in comparison with a threshold. It dynamically adjusts this threshold on every block insertion. Using a DRRIP [56] framework, Baek et al. [55] proposed a size-aware insertion policy that gives the big-size cache blocks a higher chance of eviction. On evictions, it also chooses the biggest-size cache block in case there were multiple possible victims. Employing these policies, ECM has the potential to improve effective capacity, cache miss rate, and overall system performance. Pekhimenko et al. [54] similarly propose tailored replacement and insertion policies for compressed caches.

Interactions with prefetching: Alameldeen and Wood [57] show that compression and prefetching can interact in strong positive ways. Prefetching, in general, suffers from bandwidth pressure and cache pollution, while compression can alleviate both of these. Similarly, prefetching can help compression by hiding the decompression latency. Alameldeen and Wood [57] propose an adaptive prefetching mechanism that enables prefetching whenever beneficial. They use extra tags already provided for fitting more compressed blocks to also detect useless and harmful prefetches. In their compressed cache, they double the number of tags to potentially track twice the number of compressed blocks. However, in many cases, not all the blocks are compressible, so there are extra tags not being used. They leverage these tags to track recently evicted blocks and to find whether prefetched blocks were evicting useful ones. They use a saturating counter that they incremented on useful prefetches, and decremented on useless or harmful prefetches. Using this counter, they disabled prefetching when it did not help. Overall, they show that by leveraging the interaction between compression and prefetching, they can significantly improve performance.

3.4   CACHE COMPRESSION TO IMPROVE CACHE POWER AND AREA

In addition to adopting compression to improve cache effective capacity, some techniques aim at reducing cache power and area using compression. In general, in compressed caches, we can reduce cache power on a block access if the dissipated power to compress/decompress the block is lower than the dissipated power to access the uncompressed block minus the dissipated power to access the compressed block. Thus, in all these techniques, they use simple compression algorithms (such as significance-based) with small power overheads at the cost of lower compressibility compared to LZ-based compression algorithms.

Yang et al. [40] exploit frequent value locality to improve cache dynamic power. They compress a cache line into half, if possible, otherwise, stored it as uncompressed. They partitioned the cache data array into two sub-arrays such that on an access to a compressed block (i.e., a frequent value), they would only activate the first data sub-array. Otherwise, it would require an additional cycle to access the second data sub-array. In this way, they could reduce cache dynamic energy consumption for frequent value accesses, at the cost of higher access time for non-frequent value accesses.

Significance-compression [45] similarly improves cache power by accessing half of a cache block if compressed, and packing more blocks in the cache. Dynamic zero compression (DZC) [52] also reduces the L1 cache dynamic power by storing and accessing only non-zero bytes of a block in the data array.

In a recent work, Kim et al. [53] aim at reducing the L2 cache area and power in single processor embedded systems. The proposal halved the L2 cache size, and compressed cache blocks and stored them in half size in the L2 cache. If a block was not compressible, it stored its first half in the L2 cache, and its second half in a small cache, called the residue cache. By reducing the size of the L2 cache and accessing half-sized blocks, this technique reduced both area and power.

3.5   SUMMARY

This chapter explains multiple compressed cache mechanisms. Several proposals have shown effectiveness of compression in improving effective cache capacity, reducing cache area, and reducing cache power. In this chapter, we derived a taxonomy of these proposals. We classify previous work using three main design factors: (1) how to provide the additional tags, (2) allocation granularity of compressed blocks, and (3) how to find the corresponding block given a matching tag. Most proposals trade some of these factors for another for different reasons, including simplifying the design complexity, or reducing wasted space. Overall, compressed caching has shown to be effective in improving cache utilization. We also presented several policies proposed to improve efficiency of compressed caches.

CHAPTER 4

Memory Compression

While data compression can increase the utilization of caches and make better use of on-chip resources, applying it to main memory can potentially increase main memory capacity substantially at a low cost. This is important for many reasons. First of all, in server installations a substantial fraction of the infrastructure and ownership cost is due to main memory. Regarding cost of ownership, main memory consumes a lot of power, both dynamically as well as statically. Dynamically, a main memory access is several times more power consuming than an on-chip cache access. In addition, DRAM is volatile and needs to be refreshed. The refresh rate has steadily increased with the capacity of main memory, making it a significant component of the dynamic power consumption. As for static power consumption, it is proportional to the main memory capacity, which has also increased. Altogether, main memory accounts for a significant fraction of a system's overall power consumption. Apart from component and cost of ownership of servers, other systems, and in particular mobile computers, are also sensitive to memory cost. In addition, they are also form-factor constrained; even if more memory were affordable, from a cost perspective, it may not be possible to fit more memory into a computerized device, e.g., a smart phone or a tablet.

Data compression can squeeze more data into available main memory. This can result in lower cost per stored byte, lower power consumed per byte, reduced size per stored byte, fewer accesses to disks, and fewer context switches, all of which are potentially important values for a wide range of computerized products. Despite these advantages, the design space of data compression applied to main memory is only partially explored. In this chapter, we review the state of the art of this technology, with an emphasis on hardware-assisted techniques.

Early work on memory compression aimed at increasing the memory capacity by compressing memory pages selected to be written back to disk. The freed-up space made it possible for victimized pages to stay longer in memory, thus improving performance. Douglis [66] took exactly that approach and demonstrated that one could get a performance boost by using a software-implemented compression algorithm since its decompression latency is substantially lower than the disk access time. Wilson et al. [121] focused on the same use case and contributed with new compression algorithms that demonstrated significant gains. Their investigations led to commercial uptake in commercially used operating systems (OS)—e.g., in Apple OSX, Linux, and recently Windows—that today manage the swap space using memory compression.

Ultimately, one would like to compress all pages in memory, not only those pages selected for eviction, in order to take full advantage of memory compression. In Section 4.1, we introduce a baseline system architecture as a framework for how to implement a compressed memory system representing the state of the art. We use this to identify the design issues in the design space of compressed memory systems by focusing on compression algorithms in Section 4.2 and compaction strategies in Section 4.3. Finally, in Section 4.4, we summarize.

4.1   BASELINE SYSTEM ARCHITECTURE OF A COMPRESSED MEMORY SYSTEM

Figure 4.1 shows a baseline multicore system—i.e., without memory compression—that we consider in this chapter. Each core (P) has private L1 and L2 caches connected to a shared L3 cache, which in turn interfaces to a memory controller to bring data from/to the off-chip DRAM memory. Virtual addresses are translated to physical addresses either before or in parallel with the L1 cache access, which means that the entire cache and memory hierarchy are addressed using physical addresses. When the operating system (OS) boots, it determines the amount of physical memory in the system, which will normally remain fixed throughout its operation.[1] In a virtualized environment, a guest OS is typically booted with the maximum possible memory size and the virtual machine monitor manages the available space, e.g., by using a "balloon" driver [128]. A balloon driver allows the virtual machine monitor to allocate and "pin" guest physical memory pages, effectively reducing the amount of main memory available to the guest OS.

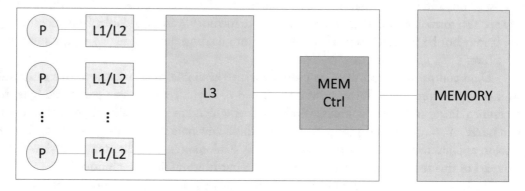

Figure 4.1: Baseline system architecture.

Now consider extending this baseline system to employ only main-memory compression. That is, where memory pages are compressed while they are stored in main memory, but decompressed when moved or copied into the cache hierarchy or to secondary storage. Like cache compression, memory compression introduces several issues involving compression and compaction. First, memory must be decompressed on each access, before being placed in the cache hierarchy or written to secondary storage. Second, the size of a memory page is no longer fixed, breaking the conventional linear translation between virtual and physical page addresses. Third, the capacity of

[1]Some enterprise systems support "hot-plug," allowing the system to dynamically grow or shrink the physical memory size.

main memory is no longer fixed, requiring the operating system to dynamically add and remove pages depending upon the compression ratio of the memory contents. Finally, processor writes may change a block's compressibility, introducing fragmentation or requiring re-compaction on cache write backs. While these issues are abstractly similar to those in cache compression, in practice they are often handled fundamentally differently due to the differences between cache and main memory design.

The first issue is how compressed memory blocks are decompressed. Like caches, the decompression latency ends up on the critical memory access path, and thus the choice of compression algorithm is critical to overall performance. However, because main memory latency is typically much longer than cache access latency, this may enable more complex and more effective compression algorithms.

The second issue is how to locate a block in a compressed memory page, given that neither pages nor blocks have a fixed size. This is more complex than the analogous problem in cache compressions for several reasons. First, main memory is typically fully associative at the page level, increasing the number of possible storage locations. Second, memory is many times larger than caches, further increasing the number of possible locations. Finally, because the cache hierarchy uses the physical address to identify which data is which, compressing a memory page cannot rename the physical address of a cache block. As a result, locating a compressed memory block typically involves translating a physical address into a separate compressed address space. Figure 4.2 illustrates this *physical-to-compressed* address translation.

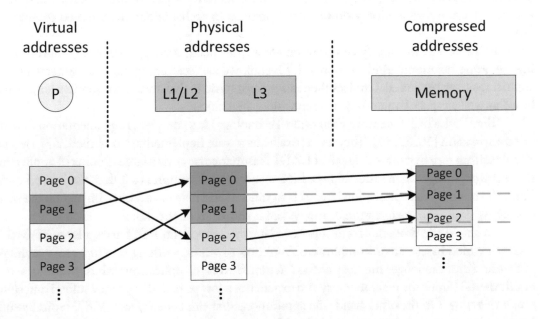

Figure 4.2: Logical address translation process in compressed memory system.

The third question is how to deal with a main memory of varying size. In cache compression, the varying size simply changes effective capacity of the cache, affecting performance but remaining transparent to software. In contrast, when the main memory size increases, the operating system must allocate new "physical" pages to take advantage of the increased capacity. Worse, when the size decreases, the operating system must free "physical" pages. This requires a solution similar to the balloon driver used by virtual machine monitors to manage the amount of main memory available to the guest OS [128].

Finally, since compressibility naturally changes during execution, memory pages and individual blocks may change in size. In particular, if a block is modified in the local cache hierarchy, its size may have changed by the time it is written back to main memory. If it is now smaller, storing it back into its original space results in internal fragmentation from the unused space. If it is larger, additional space must be found or made, perhaps requiring recompaction of blocks within a memory page or resulting external fragmentation. In either case, fragmentation inevitably reduces the effective memory capacity. Like compressed caches, keeping fragmentation low is important.

4.2   COMPRESSION ALGORITHMS

As with caches, there are two conflicting goals in choosing a compression algorithm. First, it should be effective, leading to a high compression ratio and larger effective memory capacity. Second, since decompression ends up on the critical memory access path, it should impose as little performance loss as possible. However, since memory access is much slower than cache access, one can afford modestly slower decompression in exchange for an increased compression ratio.

In Table 4.1, we classify compression algorithms being used for memory compression using the same taxonomy as in Section 2.2. Overall, we can see that so far only general-purpose algorithms have been used. Choices then range from static to dynamic and cover temporal-value-based as well as spatial-value-based compression algorithms.

The IBM MXT (Memory Expansion Technology) is a complete implementation of memory compression [30, 31, 58]. They use a parallel hardware implementation of the LZ77 Lempel-Ziv algorithm (see Section 2.3.2) called LZ1. Measurements across a large number of applications show typical compression ratios of two. LZ1 manages to decompress a 1-KB block in 64 cycles. Thus, the memory decompression latency is significant and may have a huge impact on the memory access time. The organization of the system is shown in Figure 4.3.

To shield the processor from this extra latency component, MXT uses a large 32-MB L3 cache in which blocks are uncompressed. Applications whose working sets fit in that cache will not suffer from the longer memory access latency. However, applications whose working sets do not fit the L3 but fit the main memory and experience a low page fault rate may suffer from slower memory access. On the other hand, the applications that can benefit from MXT typically suffer from a high page fault rate and are helped by more main memory capacity despite the longer access time.

Table 4.1: Compression algorithms used for memory compression

		Temporal-Value Based	Spatial-Value Based
General-Purpose	**Static**	Null [37]	—
	Dynamic	Lempel-Ziv [27][28]	FPC [20] BDI [19]
Special-Purpose	**Static**	—	—
	Dynamic	—	—

The significant memory decompression latency of the IBM MXT motivated Ekman and Stenström [59] to propose a memory compression system that would not lead to any performance degradation for applications that do not demand more memory capacity. One of their design decisions is to use simple compression algorithms that exhibit short decompression latency. They chose Frequent Pattern Compression (FPC) compression (see Section 2.3.5). Based on their compressibility studies of memory dumps, they find that FPC in fact offers a compression ratio of close to two and that this is attributed to the prevalence of zero words which is about 40% according to their study. Their study applies compression to 64-byte blocks. The good news is that FPC-encoded blocks can be decompressed in only five cycles, i.e., much faster than LZ1 used in IBM MXT.

Ekman and Stenström [59] also observe that about 30% of the memory blocks are "null," i.e., containing only words that are zero. Dusser and Seznec [60] leverage this observation using a null-block coding that represents null blocks with a single bit of metadata. Decompressing null blocks has negligible latency (effectively an AND-gate), however, the compression ratio is quite low.

Finally, Linear Compressed Memory [61], discussed further below, tries to compress all blocks to the same size using BDI [19] (see Section 2.3.6), storing blocks that don't compress to the target size as uncompressed blocks.

4.3   COMPRESSED MEMORY ORGANIZATIONS

Like caches, compressed main memories must compact variable-size blocks and pages into a fixed-size physical storage space. The taxonomy for compressed caches introduced in Section 3.1 focuses on how compressed blocks are located and compacted to minimize internal fragmenta-

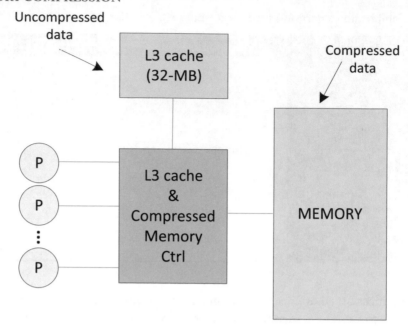

Figure 4.3: IBM MXT system organization.

tion. When applied to compressed memory, instead of mapping tags to compressed blocks, we are concerned with how addresses in the physical (uncompressed) address space are mapped to addresses in the compressed address space, what metadata is needed, and how fast the compressed block can be located. Table 4.2 classifies the systems we review using a similar taxonomy as in Section 3.1.

4.3.1 THE IBM MXT APPROACH

In the IBM MXT proposal [30, 31, 58], compressed memory is organized into 256-byte sub-blocks. The block size of the L3 cache is 1 KB. A block can in general be compressed to between zero and four sub-blocks. It will occupy zero sub-blocks if it trivially is a null block and four sub-blocks if it is uncompressed.

In order to locate the sub-blocks associated with a referenced block, the compressed memory system controller, as shown in Figure 4.3, contains a mechanism to translate physical addresses to compressed addresses, called the Compression Translation Table (CTT). CTT translations are stored in a reserved part of main memory. Each CTT entry occupies 16 bytes and can hold four sub-block addresses for the four sub-blocks potentially needed to store a 1-KB block. In the special case that the block is a null block, this fact is encoded in the CTT entry. We note that address translation can result in additional memory accesses to retrieve the information from the

Table 4.2: Compressed memory taxonomy

Phys.-to-Compr. Mapping		Data		
		Full Block	Sub-Block	Byte
Per Block	Direct One-to-One Phys/Compr. Mapping	—	MXT [30][31][58], Ekman/Stenstrom [59], LCP [61]	—
	Decoupled Forward Pointers	—	—	—
	Decoupled Back Pointers	—	—	—
Per Super Block	Direct One-to-One Phys/Compr. Mapping	—	—	—
	Decoupled Forward Pointers	DZC [60]	—	—
	Decoupled Back Pointers	—	—	—

CTT and the latency of it ends up on the critical memory access path. This design decision was considered reasonable because MXT uses a huge tertiary cache to shield that latency from the processors.

4.3.2 THE EKMAN/STENSTRÖM APPROACH

Ekman and Stenström's [59] objective is to offer negligible decompression latency and to deal effectively with fragmentation. Central to meeting this objective is the way by which the compressed memory is organized. They compress blocks at the granularity of conventional caches, in their design 64B. A compressed block can occupy a number of discrete sizes; in the system evaluated they use four sizes. Assuming a block size of 64B, a compressed block can occupy 0, 22, 44, and 64B, where no space is needed for a null block.

To minimize the performance overhead to find a compressed block, they propose to use a TLB-like structure, called the Block Size Table (BST). The BST has as many entries as the TLB, but is accessed in parallel with the L3 cache, as shown in Figure 4.4. Each BST entry keeps information about the location of each individual block within a page. The four discrete block sizes are encoded with 2 bits. Assuming a page size of 4 KB and a block size of 64 bytes, each entry needs 128 bits. The BST is accessed in parallel with the L3 cache, so no additional latency is incurred on an L3 cache miss.

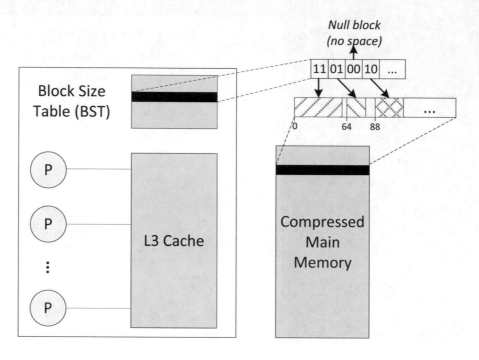

Figure 4.4: Organization of Ekman/Stenström's compressed memory system.

Fragmentation is kept low through two means. First, it uses small discrete sizes of compressed blocks. Second, when a block size changes, resulting in more or less space, moving blocks reduces fragmentation. Of course, it is important to reduce the number of block movements because these operations are costly.

A hysteresis is used such that a block/subpage must change in size by a certain quantum before block/subpage movement happens. In order to reduce the number of blocks that need to be moved, a page is divided into subpages so that only blocks inside a subpage need to be moved most of the time. Similar to assigning a hysteresis to blocks, subpages are also associated with hystereses to reduce the number of costly subpage movements.

4.3.3 THE DECOUPLED ZERO-COMPRESSION APPROACH

The motivation behind Dusser and Seznec's Decoupled Zero-Compressed (DZC) memory proposal [60] is the prevalence of null blocks. Their idea is to only compress null blocks by encoding them with a single bit. They organize the compressed memory as a decoupled super-block cache (called a "decoupled sector cache" in the original work [75]), a similar approach to the compressed DCC cache design [21–23] (see Section 3.2.3).

A super-block cache associates a set of consecutive memory blocks—a super-block—with a single address tag. In comparison with a conventional cache, the overhead of a tag is amortized across multiple blocks. Conversely, metadata such as valid and coherence states must be maintained for each block in the super-block. In a decoupled super-block cache, a block in the data array can be associated with one of several tags. An (N,P) decoupled super-block cache [75] allows any block in a super-block containing P blocks to be associated with N tags. This improves the hit rate of super-block caches, which can otherwise be very low for workloads with poor spatial locality.

In DZC, compressed memory is organized into equal size regions, called C-spaces, whose size is a multiple of the (uncompressed) page size, typically 64–512 times. A physical page is mapped to a given C-space. The idea is to encode null blocks in the metadata of a page's descriptor and allocate no other memory. Consequently, DZC stores only non-null blocks, which are stored uncompressed. Therefore, a block is either not compressed or not stored. All non-null blocks of a physical page are stored in the same C-space.

The compressed memory is managed as a set-associative decoupled super-block cache where each page is treated as a super-block and allocated within a C-space. Assuming that the size of a C-space is S and the size of a physical page is P, each non-null block of a page has S/P possible line positions in the C-space.

To locate the block in the compressed memory, each physical page is associated with a page descriptor that consists of a pointer to the C-space, where the page resides, and each block in the page is associated with an N-bit encoding whether it is a null block or not and a way-pointer to retrieve the block when it is not null. As an example, assume that S and P are 4 MB and 8 KB, respectively, and the block size is 64 bytes. Then 8KB/64B = 128 way pointers are needed, where each way pointer needs $\log_2(4MB/8KB) = 9$ bits. Using a 4-byte C-space pointer, the total size of a page descriptor is 164 bytes. DZS uses a translation cache of page descriptors alongside the memory controller (see Figure 4.1) allowing L3 misses and write back requests to locate where the block resides in the compressed memory. On an access to a null block, the translation cache will detect that the block is null and can immediately return the data without accessing memory.

While this proposal only compresses null blocks by not storing them explicitly in memory, compressibility is limited but still significant; it has been noticed that 30% of blocks are null [59]. The management of the compressed memory is also greatly simplified thanks to the two possible sizes—uncompressed or not stored. Therefore, there is no issue with fragmentation! However, since a C-space can only accommodate a certain number of pages, it may happen that the C-space gets exhausted and pages will have to be moved. This will cause performance losses but it is expected to happen quite rarely.

4.3.4   THE LINEAR COMPRESSED PAGES APPROACH

The proposed compressed memory organizations described so far need to locate the accessed memory block in the compressed memory through an additional address translation step between physical addresses and compressed addresses. This step involves making a lookup in a table. In MXT, this corresponds to additional costly memory accesses; in the other reviewed approaches it involves a translation cache access. In particular, in Ekman and Stenström's proposal, address translation involves address calculation whose latency is partly hidden by happening in parallel with the access to the L3 cache. Pekhimenko et al. [61] note that this can waste power as address calculations happen for all L3 accesses even if they do not miss. Another drawback of Ekman and Stenström's proposed address translation is the significant amount of metadata needed. To address these two shortcomings, Pekhimenko et al. [61] propose linear compressed pages (LCP) as a framework for realizing efficient address translation and management of compressed memory systems.

The key idea behind LCP is to restrict compressibility of pages and blocks to certain sizes. For example, assuming a baseline page size of 4 KB, LCP can restrict compression ratios to 2X, 4X, and 8X yielding compressed page sizes of 2 KB, 1 KB, and 512 B. Similarly, and key to their proposal, is that all blocks inside a page are compressed to the same size. For example, assuming a block size of 64 bytes and a compressed page size of 1 KB would require all blocks to be compressed by a factor of four down to 16 bytes. Since all blocks will not compress, the framework can handle exceptions too, that we discuss below.

The primary advantage of applying the same compression ratio to all blocks is that it greatly simplifies address calculation. The metadata needed is essentially an index into the location where the compressed page starts and the size of the compressed page. From that information, the location where each individual block resides can be calculated efficiently by a simple shift operation using the block's compression ratio as input. In addition, the amount of metadata is greatly reduced.

The challenge is to deal with blocks that do not compress as well as the predetermined compression factor. LCP deals with them as exceptions. They are not compressed and are stored in a special designated space after all the compressed blocks. Exception blocks still use up the same space as compressed block but that space is used to keep a pointer to where they reside. As a result, locating exception blocks requires additional address calculations compared to blocks that can be compressed to the predetermined ratio. LCP will obviously suffer from internal fragmentation but has been shown to compress quite well.

4.4   SUMMARY

In order to enable compression in the memory, several issues need to be tackled. First, since decompression ends up on the critical access path, like in cache compression, compression algorithm implementations must exhibit a low latency. While IBM MXT uses a Liv-Zempel-based compression algorithm, which incurs a significant latency, they use a huge tertiary cache to shield that latency. In contrast, the other reviewed compressed memory organizations use simple compression algorithms that detect common values (e.g., null and narrow values) or exploit spatial value locality with delta encoding.

Because the size of memory pages changes dynamically, a second issue is that an extra address translation step between conventional physical addresses and compressed addresses must be introduced. This translation step appears on the critical memory access path. IBM MXT tolerates this extra latency using the huge tertiary cache in front of the memory controller. Ekman and Stenström speculatively do the address translation in parallel with the L3 cache access. DZC and LCP address this latency component specifically by simplifying address translation between physical and compressed addresses. They do so by requiring that all blocks be compressed to the same size. The drawback is a lower compression ratio.

Finally, the size of memory pages change during operation, triggered by modifications. This may lead to fragmentation. The reviewed systems deal with fragmentation in different ways. In IBM MXT, a compressed 1-KB block may use 0-4 256-byte super-blocks. This may result in suboptimal compression ratios and yield internal fragmentation. Similarly, Ekman and Stenström also allow a block to be compressed to a limited number of sizes but reduce internal fragmentation by occasionally reclaiming the unused space in the block. DZC does not suffer from fragmentation as it targets null blocks. Either a block is null in which case it does not consume any space or it is not compressed. Finally, LCP like IBM MXT uses fixed size super-blocks for compressed blocks and hence can suffer from internal fragmentation.

CHAPTER 5

Cache/Memory Link Compression

The exponential growth in the number of transistors per chip has led to a rapid increase in the number of cores on a microprocessor chip. However, the bandwidth across the chip boundary has not grown as quickly as either the transistor count or the core count. Cache compression can increase the effective capacity of the cache, and thus tends to reduce the number of cache misses and reduce bandwidth demand. Nonetheless, off-chip bandwidth can be a performance bottleneck in many systems. Techniques that can improve the effective off-chip bandwidth can easily translate into higher performance for memory-bound applications.

Data compression can better utilize the available bandwidth between the on-chip cache subsystem and main memory. We refer to this use of data compression as *link compression* as it transfers compressed data over the link that connects main memory with the microprocessor chip and its on-chip caches. In this chapter we first focus on the isolated problem of compressing data that connects an uncompressed on-chip cache to an uncompressed main memory. In this use case, data must be first compressed before being transferred from main memory to the on-chip cache subsystem and then decompressed on the chip. Similarly, when a block is evicted from the last-level cache it must be compressed before being transferred to main memory, where it is decompressed before written back to DRAM. Of course, it is possible to design systems that also store compressed data in main memory (whether they seek to compact it or not), eliminating the memory side compression and decompression logic and resulting latency. Note that such an approach requires at least one bit of metadata per memory block to identify uncompressed data.

Link compression has several desirable properties. First and most obviously, by representing the transferred information using a smaller sequence of code words, it can increase the effective link bandwidth. Because contention for a throughput bottleneck can result in quadratic queuing delays, increasing the effective bandwidth can result in significant reductions in the observed latency and hence significant performance improvements. Second, compression can reduce the time it takes to transfer a cache block across the link, potentially reducing the uncontended cache miss penalty. However, in the isolated design we consider here, the compression and decompression latencies may out-weigh the reduction in transfer latency. Hence, like in cache and memory compression, the choice of compression algorithms and their implementation is critical. Finally, by reducing the amount of data transferred across the link, the transfer itself will be more energy efficient. The energy used by the compression and decompression logic must also be considered, but in general communication (e.g., link) energy tends to dominate computation energy. Thus, link compression has the potential to improve both performance and energy efficiency. In this chapter, we survey a number of techniques proposed in prior art to compress data on the cache/memory link.

Thuresson et al. [114] and Arelakis and Stenström [113] observe that link compression can exploit both spatial-value and temporal-value locality, and further decompose spatial-value locality into narrow-value (or significance) and clustered-value locality. Recall that clustered-value locality refers to the property that values tend to be similar, and thus can be efficiently encoded as the difference, or delta, from a base value. Narrow-value locality refers to the special case of clustered-value locality where the base value is zero, and thus captures the common case of small integers.

In the following, we discuss proposed link compression methods according to the classification of narrow, clustered, and temporal value locality in Sections 5.1–5.3, respectively. These techniques assume that data are compressed before being transferred and decompressed after the transfer. If data are already stored compressed at memory, the compression step can be avoided. In Section 5.4, we review techniques that utilize that data is stored compressed in memory. Finally, in Section 5.5, we summarize.

5.1   LINK COMPRESSION FOR NARROW VALUE LOCALITY

Many studies have shown that narrow value locality is prevalent, meaning that integers that can be encoded by typically far fewer bits than the integer data type allows are common. Each value is encoded into two parts: the fixed-width prefix that indicates the size needed by the value and the actual value. In general, given an N-bit value, whose M most significant bits are all zeros or ones (i.e., replicating the sign-bit), one can encode it using $\log_2(N) + N - M + 1$ bits. Note that the extra bit is required to encode the sign bit. A 32-bit value therefore consists of a 5-bit prefix, that encodes the number of bits needed to store the value, plus the actual value bits. As an example, the value 0x00000079 is encoded as 0x0879, yielding a compression factor of 32/13.

Frequent Pattern Compression (FPC) [20], introduced in Section 2.3.5, uses a variation of this general approach, but rather than allow all possible value sizes it compresses only 4-bit, 8-bit, and 16-bit values where the remaining bits are sign extensions. FPC is a fast and simple approach to compress narrow integers and requires no state, as many other compression algorithms do. As a result, it is possible to apply FPC to all words in a block in parallel. The compression latency is low and is essentially the delay through a priority encoder. Applying FPC to larger numbers can, however, result in a significant overhead in terms of the prefix needed to code the sign-extension bits. It is possible to reduce the overhead by partitioning the value space into a small set of partitions. For example, the value space could be partitioned into four groups of values that need 32, 24, 16, and 8 bits. This would then need only a 2-bit prefix code. Thuresson et al. [114] analyze the compressibility of this scheme using the SPEC 2000 benchmarks, media, and server applications and observe that 35% more bandwidth can be gained.

5.2   LINK COMPRESSION FOR CLUSTERED VALUE LOCALITY

Clustered value locality refers to the property that it is common that values are clustered around a base value. This property holds especially for addresses. Hammerstrom and Davidson [126] introduce *delta encoding* to exploit address value locality. The idea is to dynamically find one base value for each such value cluster. For all values that belong to a cluster, one then only sends the difference or delta between the actual value and the base value.

As shown in Figure 5.1, this scheme is implemented using a value cache on both sides of the link that caches the same set of base-value candidates. When a value is transferred, a fully associative search of all the entries in the value cache is conducted and the closest base value—the cluster value—is located. The difference between a data value and the cluster value, i.e., the delta, is transmitted along with the index to the cluster value. If the delta is larger than a given threshold, one of the cluster values will be replaced with the new value using for example LRU. This will ensure that cluster values are gradually updated to adapt to new clusters of values. It is assumed that the value caches at both sides of the link are consistent. To maintain consistency, a replacement of a cluster value at one side must be reflected on the other side.

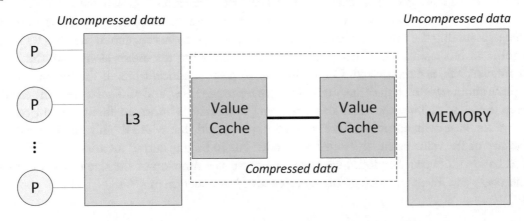

Figure 5.1: Value cache organization for cache/memory link compression.

There is a delicate trade-off between the size and compressibility of the value cache. While a larger value cache will contain more cluster values and can yield a higher compression ratio due to smaller deltas, the number of index bits that are also transferred will be higher. In other words, as the size of the value cache increases, the compression ratio is expected to follow a U-shaped curve. Another consideration is the compression latency, which increases with the size of the cache. Thuresson et al. [114] analyzed this scheme and found that bandwidth can be reduced by 60% with a 32-entry value cache.

5.3 LINK COMPRESSION FOR TEMPORAL VALUE LOCALITY

Instead of looking for clusters in the value space, an alternative is to look for locality with respect to distinct values, i.e., temporal value locality. In this section, we consider two schemes that go after distinct values. Like the delta encoding scheme, the first scheme proposed by Farrens and Park [43] for address line compaction, and extended by Citron and Rudolph for data transfers [44], referred to as the *Citron scheme*, identifies temporal value locality with respect to fixed regions in the value space by only inspecting the 16 most significant bits. The second scheme exploits temporal value locality at the granularity of individual values in the 32-bit value space. This scheme is proposed by Yang et al. [122] and is referred to as *Frequent Value Encoding* (FVE).

5.3.1 THE CITRON SCHEME

The infrastructure to implement the Citron and FVE schemes is the same as in delta encoding. It uses a pair of value caches, as shown in Figure 5.1, on each side of the link that is kept consistent. However, the management of these value caches differs for each of the schemes.

In the Citron scheme, the value cache contains the 16 most significant bits of a set of values, assuming the full values have 32 bits. A value that is to be transferred matches its 16 most significant bits against all values in the value cache. On a hit, the index, along with the 16 least significant bits, are transferred. On a miss, the 16 most significant bits of the transferred value replace another value in the value cache, using for example LRU, and all 32 bits are transferred. To keep the value cache at the other side consistent, that cache is informed about the replacement.

As with delta encoding, Thuresson et al. [114] find that while the hit rate increases with the size of the value cache, the number of index bits to be transferred increases too. This means that bandwidth reduction follows a U-shaped curve as a function of the size of the value cache and peaks at a 32-entry value cache with a bandwidth reduction of 40%.

5.3.2 FREQUENT VALUE ENCODING

The FVE scheme [122] simply keeps the full 32-bit value in the value cache unlike in the Citron scheme where only the 16 most significant bits are stored. This will save a lot of bandwidth if a limited set of distinct values is common, as only the index bits have to be transferred on a hit. On a miss, however, all 32 bits have to be sent like in the Citron scheme.

Not very surprisingly, Thuresson et al. [114] find that compressibility again follows the same trend as for Citron, and bandwidth reduction follows a U-shaped curve as a function of the size of the value cache and peaks at a 32-entry value cache. However, this scheme can save, as much as, 70% bandwidth.

5.4   LINK COMPRESSION METHODS APPLIED TO COMPRESSED MEMORY DATA

The link-compression methods discussed so far assume that the compressor/decompressor mechanisms, in essence the value caches, sit at each side of the link. When these methods were proposed, the memory controller (MC) was typically not integrated on the microprocessor chip. Then, one compressor/decompressor unit could be integrated on the microprocessor chip and the other could be integrated in the MC to enjoy improved bandwidth across the microprocessor chip boundary. Today, however, most microprocessors integrate the MC on the chip. The MC controls the DRAM directly by a well-defined data-channel protocol adopting standards like DDR3. It is not reasonable to integrate link compression mechanisms on the memory channel as they will have a non-trivial impact on timing.

To understand how we can save bandwidth on the memory channel through compression, let's review what happens on a memory access. An MC handles a 64-bit DDR3 memory channel that supports 1–4 ranks (a number of DRAM chips) that together supply the 64 bits. Assuming a request for a memory block of 64 bytes in the same DRAM row, it will generate a burst of eight channel requests, where each request will read from a distinct column.

The approach taken in MemZip [65] is to keep the data in each memory block compressed but only for the purpose of saving bandwidth, not for the purpose of gaining more memory capacity. This has the desirable property that memory blocks are easily located with the physical address and one does not have to deal with the additional address translation step between physical and compressed addresses as we discussed in Section 4.3. Instead, the leverage of a compressed memory block is to save bandwidth by generating fewer channel requests when a memory block is requested from the memory controller. This can translate into performance improvements and energy savings.

MemZip uses rank subsetting to generate fewer requests for blocks that compress. To do that, the MC must know the size of the compressed block to be fetched. Therefore, metadata that encodes the size of each memory block must be available to the MC. MemZip associates four bits of metadata with each memory block leading to 512 bits—the size of a memory block—of metadata for an 8-KB memory page. The metadata is stored in physical memory but cached in the memory controller in a metadata cache. The size of the metadata for all blocks in a page is the same as a 64-byte cache block. MemZip uses 127 blocks for (compressed) data and the 128th block for metadata. The DRAM row is also 8 KB meaning that the data as well as the metadata is accommodated in one row. This leads to the desirable property that accesses to the metadata will likely result in a row hit.

Since a block that is compressed does not use up all of the space in the memory block frame, one can use parts of it for metadata for other purposes. MemZip allows individual blocks to be compressed with different algorithms and encodes in the block frame which compression algorithm is used. For energy saving purposes, MemZip also proposes to use a technique called data bus inversion (DBI). The idea behind DBI is the following. Assume we want to transfer two four-bit words, say, 0000 and 1111. By allowing us to invert the bits of one four-bit word, say the second one, we will save dynamic energy by avoiding the switching of the four bits. This would cost an extra bit that encodes whether DBI is activated or not. In the example, we would send 00000 followed by 10000, where the most significant bit designates whether DBI is activated or not. MemZip demonstrates significant performance improvement and energy savings with this technique.

Finally, Sathish et al. [111] also aim at bandwidth reduction through compression but in the context of GPUs. They integrate compression and decompression in the GPU memory controller. When data is transferred between the CPU and GPU memory subsystems, it is compressed, and when the GPU accesses the data, the MC will decompress it. Like in MemZip, metadata must be accessible by the GPU MC. They create the metadata when data is compressed and store it in a look-up table. Due to the sheer size of the metadata, they store it in the GPU DRAM and use a metadata cache to access it efficiently. Interestingly, they also propose to use a lossy compression method for floating-point data that simply truncates the least significant bits of the mantissa and then apply compression. They notice that the truncation does not have any noticeable impact on the result but improves compressibility quite significantly.

5.5   SUMMARY

Link compression, as we discussed in this chapter, is yet another application of data compression in the memory hierarchy where the goal is to free up precious memory bandwidth. If data is neither compressed in memory nor in the on-chip caches, data must be compressed before being transferred and decompressed after the transfer. We reviewed a general concept to accommodate link compression, which uses value caches on both sides of the link. One reviewed scheme exploits spatial (or clustered) value locality in the memory access stream meaning that values differ by just a small amount, or delta. The value caches then contain a set of base values and when a value is transferred, the delta to the closest base value is chosen along with the index to the base value so it can be looked up at the other side. In order to exploit temporal value locality we could also keep common values (or parts of values) in the value caches. A value to be transferred does a value-cache lookup and on a hit, which means that it matches any of the values, the index of that value is sent. The two value caches sitting at each side of the link must of course be consistent for the scheme to work. Hence, when one cache replaces a value, the other cache must be notified and do the same.

Placing value caches at each side of the memory link is not practical in a memory hierarchy because of the strict timing requirements of modern memory protocols. Instead, it has been proposed to keep memory compressed at memory so as to transfer only the compressed part of a cache block. This is done in MemZip, which uses rank subsetting to only trigger partial transfers instead of transferring the entire block. Similar approaches have been proposed for memory in GPUs as well. In both cases, the memory controller must store metadata about the size of compressed blocks.

An ultimate goal would be to allow data to be compressed in memory as well as in on-chip caches. Such an end-to-end compression design eliminates the need for separate compression and decompression operations before data is sent on or received from a link, respectively. In other words, one would get link compression for free.

CHAPTER 6

Concluding Remarks

Data compression has been a subject of academic research and industrial development for decades, and has been widely deployed in long-haul communication networks and archival storage systems for many years. Conversely, it has only relatively recently been considered a reasonable possibility in the processor memory hierarchy. Over the past decade, operating systems and a few commercial hardware implementations have begun using compression to increase the capacity of main memory. With the semiconductor scaling trends that continue to widen the gap between processor performance and memory access latency and bandwidth, the potential for compression to increase the effective cache capacity, improve off-chip bandwidth, and increase memory capacity has significant commercial appeal.

In this book, we have surveyed the myriad of compression algorithms and compaction techniques that have been developed over the past two decades and sought to highlight and taxonomize the common themes that run through these designs. We hope this book serves as an introduction to the field and a catalyst to spur new research ideas.

A number of challenges remain to be addressed in future work. Most prior work focuses on one level of the memory hierarchy: L1 caches, L2 caches, last-level caches, main memory, or the links that connect them. Yet in the long run, end-to-end compression—from memory to the processor—is likely to be the right answer. Future work should explore this long-run vision, to refute it if it is wrong, or demonstrate ways to bring it to reality.

Data compression works well for integer-centric applications, which includes many commercial and server workloads, yet instructions and floating-point data still remain challenging. Thus, future work to develop compression techniques with broader utility would increase the commercial appeal of applying compression in the memory hierarchy.

Finally, cost is often the key driver behind a successful technology and thus has the greatest potential to motivate the adoption of cache compression. Much published research focuses on the ability of compression to *increase* the effective cache capacity of a given cache, yet the greatest commercial appeal is likely to be the ability to *decrease* the area required for a given effective capacity.

We look forward to the next decade of research in applying compression to the memory hierarchy and speculate that it is only a matter of time before compression becomes a core topic in graduate computer architecture courses.

References

[1] C. Lefurgy, P. Bird, I. Chen, and T. Mudge. Improving code density using compression techniques. In *Proceedings of the 30th annual ACM/IEEE International Symposium on Microarchitecture*, pp. 194–203, December 01–03, Research Triangle Park, North Carolina, 1997. DOI: 10.1109/MICRO.1997.645810. 16

[2] Y.-L. Jeang, J.-W. Hsieh, and Y.-Z. Lin. An Efficient Instruction Compression/Decompression System Based on Field Partitioning. *2005 IEEE International Midwest Symposium on Circuits and Systems*, Aug. 7–10, 2005. DOI: 10.1109/MWSCAS.2005.1594495.

[3] S. Chandar, M. Mehendale, and R. Govindarajan. Area and Power Reduction of Embedded DSP Systems using Instruction Compression and Re-configurable Encoding. *Journal of VLSI Signal Processing Systems*, v. 44 n. 3, pp. 245–267, September 2006. DOI: 10.1007/s11265-006-8538-6.

[4] I. Chen, P. Bird, and T. Mudge. The impact of instruction compression on I-cache performance. *Tech. Rep. CSE-TR-330–97*, EECS Department, University of Michigan, 1997.

[5] L. Benini, A. Macii, E. Macii, and M. Poncino. Selective instruction compression for memory energy reduction in embedded systems. In *Proceedings IEEE International Symposium Low-Power Electronics and Design*, pp. 206–211, 1999. DOI: 10.1145/313817.313927. 16

[6] K.D. Cooper and N. McIntosh. Enhanced code compression for embedded RISC processors. In *Proceedings of the ACM SIGPLAN 1999 conference on Programming language design and implementation*, pp. 139–149, May 01–04, Atlanta, Georgia, 1999. DOI: 10.1145/301618.301655. 16

[7] A. Wolfe and A. Chanin. Executing compressed programs on an embedded RISC architecture. In *Proceedings of the 25th Annual International symposium on Microarchitecture (MICRO 25)*, 1992. DOI: 10.1109/MICRO.1992.697002. 16

[8] S.Y. Larin and T.M. Conte. Compiler-driven cached code compression schemes for embedded ILP processors. In *MICRO 32: Proceedings 32nd Annual International Symposium on Microarchitecture*, IEEE, pp. 82–92, 1999. DOI: 10.1109/MICRO.1999.809446.

[9] H. Lekatsas and W. Wolf. SAMC: a code compression algorithm for embedded processors. In *IEEE Trans.: Comp.-Aided Des. Integ. Cir. Sys.*, vol. 18, Issue:12, Nov., pp. 1689–1701, 2006. DOI: 10.1109/43.811316.

[10] M. Thuresson and P. Stenström. Evaluation of extended dictionary-based static code compression schemes. *Conf. Computing Frontiers*, pp. 77–86, 2005. DOI: 10.1145/1062261.1062278. 16

[11] M. Thuresson, M. Själander, and P. Stenström. A Flexible Code Compression Scheme Using Partitioned Look-Up Tables. In *HiPEAC '09 Proceedings of the 4th International Conference on High Performance Embedded Architectures and Compilers*, 2009. DOI: 10.1007/978-3-540-92990-1_9. 16

[12] M. Isenburg, P. Lindstrom, and J. Snoeyink. Lossless Compression of Floating-Point Geometry. In *Proceedings of CAD'3D*, May 2004. DOI: 10.1016/j.cad.2004.09.015. 17

[13] M. Isenburg, P. Lindstrom, and J. Snoeyink. Lossless Compression of Predicted Floating-Point Geometry. *Computer-Aided Design*, Vol. 37, Issue 8, pp. 869–877, July 2005. DOI: 10.1016/j.cad.2004.09.015. 17

[14] P. Lindstrom and M. Isenburg. Fast and Efficient Compression of Floating-Point Data. In *IEEE Transactions on Visualization and Computer Graphics, Proceedings of Visualization*, 12(5), pp. 1245–1250, September-October 2006. DOI: 10.1109/TVCG.2006.143. 17

[15] P. Ratanaworabhan, J. Ke, and M. Burtscher. Fast Lossless Compression of Scientific Floating-Point Data. *Proc. Data Compression Conf. (DCC'06)*, pp. 133–142, Mar. 2006. DOI: 10.1109/DCC.2006.35. 5, 17

[16] M. Burtscher and P. Ratanaworabhan. FPC: A High-Speed Compressor for Double-Precision Floating-Point Data. *IEEE Transactions on Computers*, vol. 58, no. 1, pp. 18–31, Jan. 2009. DOI: 10.1109/TC.2008.131. 17

[17] A. Beers, M. Agrawala, and N. Chaddha. Rendering from Compressed Textures. *Computer Graphics, Proc. SIGGRAPH'96*: 373–378, 1996. DOI: 10.1145/237170.237276. 6

[18] X. Chen, L. Yang, R.P. Dick, L. Shang, and H. Lekatsas. C-pack: a high-performance microprocessor cache compression algorithm. In *IEEE Transactions on VLSI Systems*, 2010. DOI: 10.1109/TVLSI.2009.2020989. 4, 5, 13

[19] G. Pekhimenko, V. Seshadri, O. Mutlu, P.B. Gibbons, M.A. Kozuch, and T.C. Mowry. Base-delta-immediate compression: practical data compression for on-chip caches. In *Proceedings of the 21st International Conference on Parallel Architectures and Compilation Techniques (PACT '12)*, ACM, New York, pp. 377–388, 2012. DOI: 10.1145/2370816.2370870. xi, 5, 7, 12, 18, 19, 37

[20] A. Alameldeen and D. Wood. Adaptive Cache Compression for High-Performance Processors. In *Proceedings of the 31st Annual International Symposium on Computer Architecture*, 2004. DOI: 10.1145/1028176.1006719. 5, 19, 22, 24, 27, 30, 46

[21] S. Sardashti and D. Wood. Decoupled Compressed Cache: Exploiting Spatial Locality for Energy-Optimized Compressed Caching. *Annual IEEE/ACM International Symposium on Microarchitecture*, 2013. DOI: 10.1145/2540708.2540715. 22, 26, 29, 41

[22] S. Sardashti and D. Wood. Decoupled Compressed Cache: Exploiting Spatial Locality for Energy Optimization. *IEEE Micro Top Picks from the 2013 Computer Architecture Conferences*, 2013. DOI: 10.1109/MM.2014.42. 22, 26

[23] S. Sardashti, A. Seznec, and D. Wood. Skewed Compressed Caches. In *47th Annual IEEE/ACM International Symposium on Microarchitecture (MICRO-47)*, 2014. DOI: 10.1109/MICRO.2014.41. 4, 22, 27, 41

[24] D.A. Huffman. A Method for the Construction of Minimum-Redundancy Codes. In *Proc. Inst. Radio Engineers*, 40(9):1098–1101, September 1952. DOI: 10.1109/JR-PROC.1952.273898. 5, 9

[25] J.S. Vitter. Design and Analysis of Dynamic Huffman Codes. *Journal of the ACM*, 34(4):825–845, October 1987. DOI: 10.1145/31846.42227. 9

[26] I.H. Witten, R.M. Neal, and J.G. Cleary. Arithmetic Coding for Data Compression. *Communications of the ACM*, 30(6):520–540, June 1987. DOI: 10.1145/214762.214771. 5

[27] J. Ziv and A. Lempel. A Universal Algorithm for Sequential Data Compression. *IEEE Transactions on Information Theory*, 23(3):337–343, May 1977. DOI: 10.1109/TIT.1977.1055714. 4, 8

[28] J. Ziv and A. Lempel. Compression of Individual Sequences Via Variable-Rate Coding. *IEEE Transactions on Information Theory*, 24(5):530 –536, September 1978. DOI: 10.1109/TIT.1978.1055934. 4, 8

[29] D.A. Lelewer and D.S. Hirschberg. Data Compression. *ACM Computing Surveys*, 19(3):261–296, September 1987. DOI: 10.1145/45072.45074. 6, 10

[30] R.B. Tremaine, T.B. Smith, M. Wazlowski, D. Har, K.-K. Mak, and S. Arramreddy. Pinnacle: IBM MXT in a Memory Controller Chip. *IEEE Micro*, 21(2):56–68, March/April 2001. DOI: 10.1109/40.918003. 19, 36, 38

[31] R.B. Tremaine, P.A. Franaszek, J.T. Robinson, C.O. Schulz, T.B. Smith, M.E. Wazlowski, and P.M. Bland. IBM Memory Expansion Technology (MXT). *IBM Journal of Research and Development*, 45(2):271–285, Mar. 2001. DOI: 10.1147/rd.452.0271. 36, 38

[32] P. Franaszek, J. Robinson, and J. Thomas. Parallel Compression with Cooperative Dictionary Construction. In *Proceedings of the Data Compression Conference*, DCC'96, pp. 200–209, Mar. 1996. DOI: 10.1109/DCC.1996.488325. 25

[33] L.M. Stauffer and D.S. Hirschberg. Parallel Text Compression. *Technical Report TR91–44*, Revised, University of California, Irvine, 1993.

[34] J. Lee, M. Winslett, X. Ma, and S. Yu. Enhancing Data Migration Performance via Parallel Data Compression. In *Proceedings of the 16th International Parallel and Distributed Processing Symposium (IPDPS)*, pp. 47–54, April 2002. DOI: 10.1109/IPDPS.2002.1015528.

[35] P.A. Franaszek, P. Heidelberger, D.E. Poff, R.A. Saccone, and J.T. Robinson. Algorithms and Data Structures for Compressed-Memory Machines. *IBM Journal of Research and Development*, 45(2):245–258, Mar. 2001. DOI: 10.1147/rd.452.0245.

[36] P.A. Franaszek and J.T. Robinson. On Internal Organization in Compressed Random-Access Memories. *IBM Journal of Research and Development*, 45(2):259–270, Mar. 2001. DOI: 10.1147/rd.452.0259.

[37] J. Dusser, T. Piquet, and A. Seznec. Zero-content augmented cache. In *Proceedings of the 23rd International Conference on Supercomputing*, 2009. DOI: 10.1145/1542275.1542288. 4, 5, 7, 18

[38] J. Yang and R. Gupta. Frequent Value Locality and its Applications. *ACM Transactions on Embedded Computing Systems*, 1(1):79–105, Nov. 2002. DOI: 10.1145/581888.581894. 4, 7, 11

[39] J. Yang, Y. Zhang, and R. Gupta. Frequent Value Compression in Data Caches. In *Proceedings of the 33rd Annual IEEE/ACM International Symposium on Microarchitecture*, pp. 258–265, Dec. 2000. DOI: 10.1145/360128.360154. 23

[40] J. Yang and R. Gupta. Energy Efficient Frequent Value Data Cache Design. In *Proceedings of the 35th Annual IEEE/ACM International Symposium on Microarchitecture*, pp. 197–207, Nov. 2002. DOI: 10.1109/MICRO.2002.1176250. 32

[41] Y. Zhang, J. Yang, and R. Gupta. Frequent Value Locality and Value-centric Data Cache Design. In *Proceedings of the Ninth International Conference on Architectural Support for Programming Languages and Operating Systems*, pp. 150–159, Nov. 2000. DOI: 10.1145/378993.379235. 4

[42] K. Kant and R. Iyer. Compressibility Characteristics of Address/Data transfers in Commercial Workloads. In *Proceedings of the Fifth Workshop on Computer Architecture Evaluation Using Commercial Workloads*, pp. 59–67, Feb. 2002.

[43] M. Farrens and A. Park. Dynamic Base Register Caching: A Technique for Reducing Address Bus Width. In *Proceedings of the 18th Annual International Symposium on Computer Architecture*, pp. 128–137, May 1991. DOI: 10.1145/115953.115966. 7, 48

[44] D. Citron and L. Rudolph. Creating a Wider Bus Using Caching Techniques. In *Proceedings of the First IEEE Symposium on High-Performance Computer Architecture*, pp. 90–99, Feb. 1995. DOI: 10.1109/HPCA.1995.386552. 7, 48

[45] N.S. Kim, T. Austin, and T. Mudge. Low-Energy Data Cache Using Sign Compression and Cache Line Bisection. In *Second Annual Workshop on Memory Performance Issues (WMPI)*, in conjunction with ISCA-29, 2002. 24, 32

[46] A. Arelakis and P. Stenström. SC^2: A statistical compression cache scheme. In *Proceedings of the 41st Annual International Symposium on Computer Architecture*, 2014. DOI: 10.1145/2678373.2665696. xi, 6, 10, 18, 19, 25

[47] J.-S. Lee, W.-K. Hong, and S.-D. Kim. Design and Evaluation of a Selective Compressed Memory System. In *Proceedings of Internationl Conference on Computer Design (ICCD'99)*, pp. 184–191, Oct. 1999. DOI: 10.1109/ICCD.1999.808424. 24

[48] J.-S. Lee, W.-K. Hong, and S.-D. Kim. An On-chip Cache Compression Technique to Reduce Decompression Overhead and Design Complexity. *Journal of Systems Architecture: the EUROMICRO Journal*, 46(15):1365–1382, Dec. 2000. DOI: 10.1016/S1383-7621(00)00030-8. 24

[49] E.G. Hallnor and S.K. Reinhardt. A Fully Associative Software-Managed Cache Design. In *Proceedings of the 27th Annual International Symposium on Computer Architecture*, 2000. DOI: 10.1145/342001.339660. 25

[50] E. Hallnor and S. Reinhardt. A Unified Compressed Memory Hierarchy. In *Proceedings of the 11th International Symposium on High-Performance Computer Architecture*, 2005. DOI: 10.1109/HPCA.2005.4. 25

[51] J. Dusser, T. Piquet, and A. Seznec. Zero-content augmented caches. In *Proceedings of the 23rd international conference on Supercomputing*, 2009. DOI: 10.1145/1542275.1542288.

[52] L. Villa, M. Zhang, and K. Asanovic. Dynamic zero compression for cache energy reduction. In *Proceedings of the 33rd Annual ACM/IEEE International Symposium on Microarchitecture*, 2000. DOI: 10.1145/360128.360150. 4, 32

[53] S. Kim, J. Kim, J. Lee, and S. Hong. Residue Cache: A Low-Energy Low-Area L2 Cache Architecture via Compression and Partial Hits. In *Proceedings of the 44th Annual IEEE/ACM International Symposium on Microarchitecture*, 2011. DOI: 10.1145/2155620.2155670. 32

[54] G. Pekhimenko, T. Huberty, R. Cai, O. Mutlu, P.P. Gibbons, M.A. Kozuch, and T.C. Mowry. Exploiting Compressed Block Size as an Indicator of Future Reuse. In *Proceedings of the 21st International Symposium on High-Performance Computer Architecture (HPCA)*, Bay Area, CA, Feb. 2015. DOI: 10.1109/HPCA.2015.7056021. 31

[55] S. Baek, H.G. Lee, C. Nicopoulos, J. Lee, and J. Kim. ECM: Effective Capacity Maximizer for High-Performance Compressed Caching. In *Proceedings of IEEE Symposium on High-Performance Computer Architecture*, 2013. DOI: 10.1109/HPCA.2013.6522313. 31

[56] A. Jaleel, K.B. Theobald, S.C. Steely, Jr., and J. Emer. High performance cache replacement using re-reference interval prediction (rrip). In *Proceedings of the 37th Annual International Symposium on Computer Architecture*, ISCA '10, ACM, pp. 60–71, New York, 2010. DOI: 10.1145/1816038.1815971. 31

[57] A.R. Alameldeen and D.A. Wood. Interactions between compression and prefetching in chip multiprocessors. In *Proc. Int. Symp. High-Performance Computer Architecture*, pp. 228–239, 2007. DOI: 10.1109/HPCA.2007.346200. 7, 11, 31

[58] B. Abali, H. Franke, X. Shen, D.E. Poff, and T.B. Smith. Performance of Hardware Compressed Main Memory. In *Proceedings of the 7th IEEE Symposium on High-Performance Computer Architecture*, 2001. DOI: 10.1109/HPCA.2001.903253. 36, 38

[59] M. Ekman and P. Stenström. A robust main-memory compression scheme. *SIGARCH Computer Architecture News*, 2005. DOI: 10.1145/1080695.1069978. 4, 5, 7, 37, 39, 41

[60] J. Dusser and A. Seznec. Decoupled Zero-Compressed Memory. In *Proceedings of the 6th International Conference on High Performance and Embedded Architectures and Compilers*, 2011. DOI: 10.1145/1944862.1944876. 4, 5, 37, 40

[61] G. Pekhimenko, V. Seshadri, Y. Kim, H. Xin, O. Mutlu, P.B. Gibbons, M.A. Kozuch, and T.C. Mowry. Linearly Compressed Pages: A Low Complexity, Low-Latency Main Memory Compression Framework. *Annual IEEE/ACM International Symposium on Microarchitecture*, 2013. DOI: 10.1145/2540708.2540724. 28, 37, 42

[62] M. Kjelso, M. Gooch, and S. Jones. Design and Performance of a Main Memory Hardware Data Compressor. In *Proceedings of the 22nd EUROMICRO Conference*, 1996. DOI: 10.1109/EURMIC.1996.546466.

[63] J.L. Nunez and S. Jones. Gbit/s Lossless Data Compression Hardware. *IEEE Transactions on VLSI Systems*, 11(3):499–510, June 2003. DOI: 10.1109/TVLSI.2003.812288.

[64] Y. Zhang and R. Gupta. Data Compression Transformations for Dynamically Allocated Data Structures. In *Proceedings of the International Conference on Compiler Construction (CC)*, pp. 24–28, April 2002. DOI: 10.1007/3-540-45937-5_4.

[65] A. Shafiee, M. Taassori, R. Balasubramonian, and A. Davis. MemZip: Exploring Unconventional Benefits from Memory Compression. *HPCA*, 2014. DOI: 10.1109/HPCA.2014.6835972. 49

[66] F. Douglis. The Compression Cache: Using On-line Compression to Extend Physical Memory. In *Proceedings of 1993 Winter USENIX Conference*, pp. 519–529, Jan. 1993. 33

[67] R.S. de Castro, A.P. do Lago, and D. da Silva. Adaptive Compressed Caching: Design and Implementation. In *SBAC-PAD*, 2003. DOI: 10.1109/CAHPC.2003.1250316.

[68] S. Roy, R. Kumar, and M. Prvulovic. Improving system performance with compressed Memory. In *IPDPS '01: Proceedings of the 15th International Parallel and Distributed Processing Symposium*, p. 66, IEEE Computer Society, Washington, D.C., 2001. DOI: 10.1109/IPDPS.2001.925011.

[69] P.R. Wilson, S.F. Kaplan, and Y. Smaragdakis. The case for compressed caching in virtual memory systems. In *ATEC '99: Proceedings of the annual conference on USENIX Annual Technical Conference*, pp. 101–116, USENIX Association, Berkeley, CA, 1999.

[70] Apple OS X Mavericks. http://www.apple.com/media/us/osx/2013/docs/OSX_Mavericks_Core_Technology_Overview.pdf.

[71] R. Dennard, F. Gaensslen, H. Yu, V. Rideovt, E. Bassous, and A. Leblanc. Design of Ion-Implanted MOSFET's with Very Small Physical Dimensions. *IEEE Journal of Solid-State Circuits*, 1974. DOI: 10.1109/JSSC.1974.1050511.

[72] S. Keckler. Life After Dennard and How I Learned to Love the Picojoule. In *Proceedings of the 44th Annual IEEE/ACM International Symposium on Microarchitecture*, 2011.

[73] A. Hartstein, V. Srinivasan, T.R. Puzak, and P.G. Emma. On the Nature of Cache Miss Behavior: Is It $\sqrt{2}$? J. *Instruction-Level Parallelism*, 2008.

[74] Intel Core i7 Processors. http://www.intel.com/products/processor/corei7/.

[75] A. Seznec. Decoupled sectored caches: Conciliating low tag implementation cost and low miss ratio. *International Symposium on Computer Architecture*, 1994. DOI: 10.1109/ISCA.1994.288133. 41

[76] A. Sodani. Race to Exascale: Challenges and Opportunities. *Intl. Symp. Microarchitecture*, Dec. 2011.

[77] G.E. Moore. Cramming More Components onto Integrated Circuits. *Electronics*, pp. 114–117, Apr. 1965. DOI: 10.1109/JPROC.1998.658762.

[78] PG&E *Data Center Best Practices Guide.*

[79] M.K. Qureshi, A. Jaleel, Y.N. Patt, S.C. Steely Jr., and J.S. Emer. Adaptive insertion policies for high performance caching. In *34th International Symposium on Computer Architecture (ISCA 2007)*, June 9–13, ACM, San Diego, California, 2007. DOI: 10.1145/1250662.1250709.

[80] S.T. Srinivasan, R.D. Ju, A.R. Lebeck, and C. Wilkerson. Locality vs. Criticality. In *Proc. ISCA-28*, pp. 132–143, 2001. DOI: 10.1145/379240.379258.

[81] J. Liptay. Structural Aspects of the System/360 Model85 Part II: The Cache. *IBM Systems Journal*, 1968. DOI: 10.1147/sj.71.0015.

[82] J. Rothman and A. Smith. The Pool of Subsectors Cache Design. *International Conference on Supercomputing*, 1999. DOI: 10.1145/305138.305156.

[83] D. Yoon, M. Jeong, and Mattan Erez. Adaptive granularity memory systems: A tradeoff between storage efficiency and throughput. In *Proceedings of the 38th Annual International Symposium on Computer Architecture*, 2011. DOI: 10.1145/2024723.2000100.

[84] D. Weiss, M. Dreesen, M. Ciraula, C. Henrion, C. Helt, R. Freese, T. Miles, A. Karegar, R. Schreiber, B. Schneller, and J. Wuu. An 8MB Level-3 Cache in 32nm SOI with Column-Select Aliasing. *Solid-State Circuits Conference Digest of Technical Papers*, 2011. DOI: 10.1109/ISSCC.2011.5746309.

[85] ITRS. International technology roadmap for semiconductors, 2010 update, 2011. http://www.itrs.net.

[86] CACTI, http://www.hpl.hp.com/research/cacti/.

[87] Calculating memory system power for DDR3. *Technical Report TN-41–01*. Micron Technology, 2007.

[88] A.R. Alameldeen, M.K. Martin, C.J. Mauer, K.E. Moore, M. Xu, M.D. Hill, D.A. Wood, and D.J. Sorin. Simulating a $2M Commercial Server on a $2K PC. *IEEE Computer*, 2003. DOI: 10.1109/MC.2003.1178046.

[89] A. Alameldeen and D. Wood. Variability in Architectural Simulations of Multi-threaded Workloads. In *Proceedings of the Ninth IEEE Symposium on High-Performance Computer Architecture*, 2003. DOI: 10.1109/HPCA.2003.1183520.

[90] C. Bienia and K. Li. PARSEC 2.0: A New Benchmark Suite for Chip-Multiprocessors. In *Workshop on Modeling, Benchmarking and Simulation*, 2009.

[91] V. Aslot, M. Domeika, R. Eigenmann, G. Gaertner, W.B. Jones, and B. Parady. SPEComp: A New Benchmark Suite for Measuring Parallel Computer Performance. In *Workshop on OpenMP Applications and Tools*, 2001. DOI: 10.1007/3-540-44587-0_1.

[92] A. Seznec. A case for two-way skewed-associative caches. In *Proc. of the 20th annual Intl. Symp. on Computer Architecture*, 1993. DOI: 10.1145/173682.165152. 29

[93] A. Seznec. Concurrent Support of Multiple Page Sizes on a Skewed Associative TLB. *IEEE Transactions on Computers*, 2004. DOI: 10.1109/TC.2004.21. 30

[94] A. Seznec and F. Bodin. Skewed-associative caches. In *Proceedings of PARLE' 93*, Munich, also available as INRIA Research Report 1655, June 1993. DOI: 10.1007/3-540-56891-3_24. 30

[95] N.R. Mahapatra, J. Liu, K. Sundaresan, S. Dangeti, and B.V. Venkatrao. A limit study on the potential of compression for improving memory system performance, power consumption, and cost. *J. Instruction-Level Parallelism*, vol. 7, pp.1–37, 2005.

[96] N.R. Mahapatra, J. Liu, K. Sundaresan, S. Dangeti, and B.V. Venkatrao. The Potential of Compression to Improve Memory System Performance, Power Consumption, and Cost. In *Proceedings of IEEE Performance, Computing and Communications Conference*, Phoenix, AZ, April 2003. DOI: 10.1109/PCCC.2003.1203717.

[97] J. Gandhi, A. Basu, M. Hill, and M. Swift. BadgerTrap: A Tool to Instrument x86–64 TLB Misses. *SIGARCH Computer Architecture News (CAN)*, 2014. DOI: 10.1145/2669594.2669599.

[98] graph500—The Graph500 List: http://www.graph500.org.

[99] M. Ferdman, A. Adileh, O. Kocberber, S. Volos, M. Alisafaee, D. Jevdjic, C. Kaynak, A. Popescu, A. Ailamaki, and B. Falsafi. Clearing the Clouds: A Study of Emerging Scale-out Workloads on Modern Hardware. In *17th International Conference on Architectural Support for Programming Languages and Operating Systems (ASPLOS)*, March 2012. DOI: 10.1145/2150976.2150982.

[100] Coremark. www.coremark.org.

[101] Cloudsuite. http://parsa.epfl.ch/cloudsuite/cloudsuite.html.

[102] A. Gutierrez, R. Dreslinski, T. Wenisch, T. Mudge, A. Saidi, C. Emmons, and N. Paver. Full-system analysis and characterization of interactive smartphone applications. In *IISWC '11*, 2011. DOI: 10.1109/IISWC.2011.6114205.

[103] M. Burtscher and P. Ratanaworabhan. High throughput compression of double-precision floating-point data. In *DCC*, 2007. DOI: 10.1109/DCC.2007.44.

[104] Y. Jin and R. Chen. Instruction cache compression for embedded systems. *Berkley Wireless Research Center, Technical Report*, 2000.

[105] C. Wu, A. Jaleel, W. Hasenplaugh , M. Martonosi , S. Steely, and J. Emer. SHiP: signature-based hit predictor for high performance caching. In *Proceedings of the 44th Annual IEEE/ACM International Symposium on Microarchitecture*, 2011. DOI: 10.1145/2155620.2155671.

[106] D. Yoon, M. Jeong, and M. Erez. Adaptive granularity memory systems: A tradeoff between storage efficiency and throughput. In *Proceedings of the 38th Annual International Symposium on Computer Architecture*, 2011. DOI: 10.1145/2024723.2000100.

[107] A. Basu, D.R. Hower, M.D. Hill, and M.M. Swift. Freshcache: Statically and dynamically exploiting dataless ways. In *ICCD*, 2013. DOI: 10.1109/ICCD.2013.6657055.

[108] J. Chang and G.S. Sohi. Cooperative caching for chip multiprocessors. In *ISCA-33*, 2006. DOI: 10.1109/ISCA.2006.17.

[109] J. Chang and G.S. Sohi. Cooperative cache partitioning for chip multiprocessors. *ICS-21*, 2007. DOI: 10.1145/1274971.1275005.

[110] J. Albericio, P. Ibáñez, V. Viñals, and J.M. Llabería. The reuse cache: downsizing the shared last-level cache. In *Proceedings of the 46th Annual IEEE/ACM International Symposium on Microarchitecture*, pp. 310–321, 2013. DOI: 10.1145/2540708.2540735.

[111] V. Sathish, M.J. Schulte, and N.S. Kim. Lossless and lossy memory I/O link compression for improving performance of GPGPU workloads. In *Proceedings of the 21st International Conference on Parallel Architectures and Compilation Techniques*, 2012. DOI: 10.1145/2370816.2370864. 50

[112] M.H. Lipasti, C.B. Wilkerson, and J.P. Shen. Value locality and load value prediction. In *Proceedings of the Seventh International Conference on Architectural Support for Programming Languages and Operating Systems* (ASPLOS VII). ACM, pp. 138–147, 1996. DOI: 10.1145/237090.237173. 3

[113] A. Arelakis and P. Stenström. A Case for a Value-Aware Cache. *IEEE Computer Architecture Letters*, vol. 13, no. 1, pp. 1–4, Jan.–June 21, 2014. DOI: 10.1109/L-CA.2012.31. 3, 4, 46

[114] M. Thuresson, L. Spracklen, and P. Stenström. Memory-Link Compression Schemes: A Value Locality Perspective, *Computers, IEEE Transactions on*, vol. 57, no. 7, pp. 916–927, July 2008. DOI: 10.1109/TC.2008.28. 5, 46, 47, 48

[115] Y. Tian, S.M. Khan, D.A. Jiménez, and G.H. Loh. Last-level cache deduplication. In *Proceedings of the 28th ACM International Conference on Supercomputing* (ICS '14), pp. 53–62, 2014. DOI: 10.1145/2597652.2597655. 4, 5, 7, 14, 15

[116] L. Benini, D. Bruni, B. Ricco, A. Macii, and E. Macii. An Adaptive Data Compression Scheme for Memory Traffic Minimization in Processor-Based Systems. In *Proceedings of the IEEE International Conference on Circuits and Systems, ICCAS-02*, pp. 866–869, May 2002. DOI: 10.1109/ISCAS.2002.1010595. 5

[117] L. Benini, D. Bruni, A. Macii, and E. Macii. Hardware-Assisted Data Compression for Energy Minimization in Systems with Embedded Processors. In *Proceedings of the IEEE 2002 Design Automation and Test in Europe*, pp. 449–453, 2002. DOI: 10.1109/DATE.2002.998312. 5

[118] D. Cheriton, A. Firoozshahian, A. Solomatnikov, J.P. Stevenson, and O. Azizi. HI-CAMP: architectural support for efficient concurrency-safe shared structured data access. In *Proceedings of the 17th International Conference on Architectural Support for Programming Languages and Operating Systems* (ASPLOS XVII). ACM, pp. 287–300, 2012. DOI: 10.1145/2150976.2151007. 5, 14, 15

[119] M. Kleanthous and Y. Sazeides. CATCH: A mechanism for dynamically detecting cache-content-duplication in instruction caches. *ACM Trans. Archit. Code Optim.*, 8:11:1–11:27, October 2011. DOI: 10.1145/2019608.2019610. 5, 14

[120] P. Pujara and A. Aggarwal. Restrictive compression techniques to increase level 1 cache capacity. In *Proceedings of the 2005 International Conference on Computer Design, ICCD '05*, pp. 327–333, Washington, D.C., 2005. DOI: 10.1109/ICCD.2005.94. 5

[121] P.R. Wilson, S.F. Kaplan, and Y. Smaragdakis. The case for compressed caching in virtual memory systems. In *Proceedings of the annual conference on USENIX Annual Technical Conference (ATEC '99)*. USENIX Association, Berkeley, CA, 8, 1999. 33

[122] J. Yang, R. Gupta, and C. Zhang. Frequent Value Encoding for Low Power Data Buses, *ACM Trans. Design Automation of Electronic Systems*, vol. 9, no.3, pp. 354–384, July 2004. DOI: 10.1145/1013948.1013953. 48

[123] D. Brooks and M. Martonosi. Dynamically exploiting narrow width operands to improve processor power and performance. In *Proceedings of IEEE Int. Symp. on High-Performance Computer Architecture*, pp. 13–22, 1999. DOI: 10.1109/HPCA.1999.744314. 5

[124] S. Biswas, B.R. de Supinski, M. Schulz, D. Franklin, T. Sherwood, and F.T. Chong. Exploiting data similarity to reduce memory footprints. In *Proceedings of the 2011 IEEE International Parallel and Distributed Processing Symposium, IPDPS '11*, pp. 152–163, Washington, D.C., 2011. DOI: 10.1109/IPDPS.2011.24. 14, 15

[125] R. Sendag, Peng-fei Chuang, and D.J. Lilja. Address Correlation: Exceeding the Limits of Locality, *Computer Architecture Letters*, vol. 2, no. 1, pp. 33, January–December 2003. DOI: 10.1109/L-CA.2003.3. 15

[126] C.E. Shannon. A mathematical theory of communication. In *Bell System Technical Journal, The*, vol. 27, no. 3, pp. 379–423, July 1948. DOI: 10.1002/j.1538-7305.1948.tb01338.x. 3, 47

[127] A. Arelakis, F. Dahlgren, and P. Stenström. HyComp: A Hybrid Cache Compression Method for Selection of Data-Type-Specific Compression Methods. In *Proceedings of the 48th Annual IEEE/ACM International Symposium on Microarchitecture (MICRO-48)*, 2015. xi, 6, 17, 18

[128] C.A. Waldspurger. Memory Resource Management in VMware ESX Server. In *Proceedings of the 2002 Symposium on Operating Systems Design and Implementation*, 2002. DOI: 10.1145/844128.844146. 34, 36

[129] A. Sodani. Race to Exascale: Opportunities and Challenges, 2011. 21

Authors' Biographies

SOMAYEH SARDASHTI

Dr. Somayeh Sardashti earned her Ph.D. degree in Computer Sciences from the University of Wisconsin-Madison. Her research interests include computer systems and architecture, high performance and energy-optimized memory hierarchies, exploiting new memory, and hardware technologies for high performance database systems. She currently works in Exadata Storage Server and Database Machine group at Oracle Corporation. She was the winner of the ACM student research competition at Grace Hopper conference in 2013. She holds an M.S. in Computer Sciences from the University of Wisconsin-Madison, another Master's degree, and a B.S. in computer engineering from the University of Tehran.

ANGELOS ARELAKIS

Dr. Angelos Arelakis earned his Ph.D. degree in Computer Science and Engineering in 2015 from Chalmers University of Technology, Sweden. He is a researcher at Chalmers University of Technology and a co-founder of ZeroPoint Technologies Corp. His research focuses on high performance computer architecture, in particular designing cache and memory hierarchies that are efficiently utilized by today's multicore systems, data compression, and reconfigurable computing. He holds an M.Sc. degree in Computer Engineering from Delft University of Technology (Netherlands) and a 5-year Engineering Diploma in Electronics and Computer Engineering from the Technical University of Crete (Greece).

PER STENSTRÖM

Per Stenström earned his Ph.D. degree in computer engineering in 1990 from Lund University, Sweden. Since 1995 has been a Professor of Computer Engineering at Chalmers University of Technology, Sweden. His research interests are devoted to high-performance computer architecture and he has made major contributions to especially high-performance memory systems. He has authored or co-authored 3 textbooks, 130 publications in international journals and conferences, and around ten patents. He is regularly serving program committees of major conferences in the computer architecture field and is an Associate Editor-in-Chief of the *Journal of Parallel and Distributed Computing* and a Senior Associate Editor of *ACM Transactions on Architecture and Code Optimization*. He co-founded the HiPEAC Network of Excellence funded by the European Commission. He has also acted as General and Program Chair for a large number of conferences including the ACM/IEEE Int. Symposium on Computer Architecture, the IEEE High-Performance Computer Architecture Symposium, the IEEE International Parallel and Distributed Processing Symposium, and the ACM International Conference on Supercomputing. He is a Member-at-Large of the ACM Council, a Fellow of the ACM and the IEEE, and a member of Academia Europaea, the Royal Swedish Academy of Engineering Sciences, and the Spanish Royal Academy of Engineering.

DAVID A. WOOD

Prof. David A. Wood is a Professor in the Computer Sciences Department at the University of Wisconsin, Madison. He also holds a courtesy appointment in the Electrical and Computer Engineering Department. Dr. Wood received a B.S. in Electrical Engineering and Computer Science (1981) and a Ph.D. in Computer Science (1990), both at the University of California, Berkeley.

Dr. Wood is an ACM Fellow (2005) and IEEE Fellow (2004), received the University of Wisconsin's H.I. Romnes Faculty Fellowship (1999) and Vilas Associate (2011), and received the National Science Foundation's Presidential Young Investigator award (1991). Dr. Wood is an Associate Editor of *ACM Transactions on Architecture and Compiler Optimization*, serves as Past Chair of ACM SIGARCH, served as Program Committee Chairman of ASPLOS-X (2002), and has served on numerous program committees. Dr. Wood is an ACM Fellow, an IEEE Fellow, and a member of the IEEE Computer Society. Dr. Wood has published over 70 technical papers and is an inventor on thirteen U.S. and international patents, several of which have been licensed to industry.

Printed in the United States
by Baker & Taylor Publisher Services